T0271057

Global Real Estate Capital Markets

This book unravels the complex mechanisms involved in global real estate capital markets, enabling the reader to understand how they have grown and evolved, how they function, what determines market pricing, and how the public and private debt and equity markets are linked to each other.

Using their extensive professional experience, the authors combine a structured, rigorous understanding of the theory and academic evidence behind the main concepts with practical examples, applications, case studies, quizzes and online resources. The book will enable readers to understand for example:

- Why share prices of real estate companies can differ dramatically from the underlying value of the assets
- The differing investment objectives of different categories of investors and how this influences share prices and corporate funding decisions
- How sell-side analysts make their recommendations
- How buy-side analysts decide which sectors, funds and stocks to allocate capital to
- And how ESG considerations are relevant to capital market pricing.

The book is designed not just for advanced real estate students, but also for global finance courses, Executive Education short courses and as a primer for new entrants to the sector. It is key reading for the following groups:

- Property professionals working for a listed company wanting to understand the relationship between their underlying business and the stock market valuation
- Real Estate Private Equity teams looking to understand the valuation disconnect between public and private markets and arbitrage the Parallel Asset Pricing model
- Equity/Multi asset/Property analysts/fund managers who need to understand the specific characteristics of real estate vs the other ten equity sectors and understand when to increase and decrease sector weightings.

Online materials for this book can be found on the Routledge Resource website at https://resourcecentre.routledge.com/books/9781032288017.

Alex Moss is Director of the Real Estate Centre at Bayes Business School (formerly Cass), and Course Director for the MSc Global Finance online course and the MSc Real Estate course. He has been involved in research and transactions in the global real estate sector for over 30 years. His career has encompassed award-winning sell-side research

(BZW), investment banking (CSFB), private equity (Apax Partners Capital) and fund management (M&G and Investec). He formed AME Capital in 2002, which developed a proprietary database and analytical tool for all listed real estate companies and real estate securities funds globally. This business was sold to Macquarie Securities in 2008, where he stayed for over three years as Head of Global Property Securities Analytics. In 2012 he formed Consilia Capital, a real estate investment advisory firm, and has developed an international institutional client base. In addition, at that stage he started publishing regularly in academic journals. As well as lecturing at Bayes he is also a visiting lecturer at UCL and Cambridge University. Alex is Chairman of the EPRA Research Committee.

Kieran Farrelly was formerly a Partner and Head of Market Research for the real estate team at the Stepstone Group. Prior to Stepstone, he was a Principal at the Townsend Group and a Director at CBRE Global Investors. His roles have encompassed portfolio management, investment underwriting across a range of conduits and global investment strategy. Previously, Kieran was a research analyst with M&G Real Estate. He holds a Ph.D. in real estate and planning from the Henley Business School at the University of Reading and a master's in economics from the University of Warwick. He has guest lectured on a variety of real estate investment finance topics at UCL, the universities of Cambridge, Oxford and Reading, and London Business School. He has also authored several published academic real estate papers and is an Honorary Visiting Fellow at Bayes Business School.

Global Real Estate Capital Markets

Theory and Practice

Alex Moss and Kieran Farrelly

Routledge
Taylor & Francis Group

LONDON AND NEW YORK

Designed cover image: © Getty Images

First published 2025
by Routledge
4 Park Square, Milton Park, Abingdon, Oxon OX14 4RN

and by Routledge
605 Third Avenue, New York, NY 10158

Routledge is an imprint of the Taylor & Francis Group, an informa business

British Library Cataloguing-in-Publication Data
A catalogue record for this book is available from the British Library

ISBN: 978-1-032-28801-7 (hbk)
ISBN: 978-1-032-28800-0 (pbk)
ISBN: 978-1-003-29856-4 (ebk)

DOI: 10.1201/9781003298564

Typeset in Times New Roman
by MPS Limited, Dehradun

Access the Support Material: https://resourcecentre.routledge.com/books/9781032288017

Contents

Acknowledgements

The authors would like to acknowledge the help and feedback they have received from colleagues and students in preparing the material for this book. The greatest input from students has been from those taught on the MSc Real Estate, MSc Real Estate Investment, MSc Global Finance online and Online MBA courses at Bayes Business School, the MSc in International Real Estate and Planning at University College London, the Masters in Real Estate (flexible) at Henley Business School and the MSt in Real Estate at Cambridge University. In terms of executive education there has been very relevant feedback from clients of Bayes Executive Education and in particular participants on the six-week online "Understanding the REIT Price" course run with EPRA. We would also like to thank Stephen Yorke, founder of the Property Chronicle (www.propertychronicle) for the opportunity to reach an audience of tens of thousands of practitioners, and EPRA for their continued support and endless contributions to this field of study. Finally, we would like to thank the many attendees and respondents at the European Real Estate Society and American Real Estate Society conferences over the last decade for their comments.

The authors

The authors have an impressive pedigree in academic research within real estate, as well as practical experience in implementing these ideas globally across both buy and sell-side operations, and public and private markets. This book utilises material that has been used by them in commercial practice, and on post-graduate and executive education courses over the last ten years. They have produced many published academic papers and won an award for their paper on the performance of a blended real estate portfolio for UK DC investors.

Alex Moss is director of the Real Estate Centre at Bayes Business School (formerly Cass), and course director for the MSc Global Finance online course and the MSc Real Estate course. He has been involved in research and transactions in the global real estate sector for over 30 years. His career has encompassed award-winning sell-side research (BZW), investment banking (CSFB), private equity (Apax Partners Capital) and fund management (M&G and Investec). He formed AME Capital in 2002. This company developed a proprietary database and analytical tool for all listed real estate companies and real estate securities funds globally. In 2008, the business was sold to Macquarie Securities, where Alex stayed for over three years as Head of Global Property Securities Analytics. In 2012, he formed Consilia Capital, a real estate investment advisory firm, and developed an international institutional client base. In addition, at that stage, he started publishing regularly in academic journals. As well as lecturing at Bayes, he is also a visiting lecturer at University College London, and Cambridge University. Alex is also Chairman of the EPRA Research Committee and a member of the EPRA Advisory Board.

Kieran Farrelly is Head of Global Solutions, Real Estate at Schroders Capital and is a member of the real estate management team. He was formerly a Partner in the real estate team at the Stepstone Group and a Principal at the Townsend Group. His roles have encompassed portfolio management, investment underwriting across a range of conduits and global research and investment strategy. Kieran holds a PhD in Real Estate & Planning from Henley Business School at the University of Reading and an MSc in Economics from the University of Warwick. He has also authored several published academic real estate papers and is an Honorary Visiting Fellow at Bayes Business School.

A user guide

Real estate capital markets have grown significantly over the last ten years after recovering from the GFC. There are currently no comprehensive textbooks available which focus purely on the capital markets aspect (i.e., not valuation or economics) of real estate. Our approach is to combine a structured, rigorous understanding of the theory and academic evidence behind the main concepts with practical examples and applications, utilising the authors' own extensive experience.

The book is designed for real estate students, both under and post-graduates, but also will be used as the main textbook for global finance courses, for Executive Education short courses, and as a primer for new entrants to the sector.

For teachers, it should be noted that the book is split into three sections, each of which can be used for individual modules, namely:

Global Real Estate Markets
Private Real Estate Markets
Public Real Estate Markets

The book is also aimed at the following groups who are not in full-time tertiary education:

1 Property professionals working for a public company wanting to understand the relationship between their underlying business and the stock market valuation.
2 Real estate private equity teams looking to understand the valuation disconnect between public and private markets and to arbitrage the parallel asset pricing model.
3 Equity/multi-asset/property analysts/fund managers who need to understand the specific characteristics of real estate vs. the other 10 equity sectors and understand when to increase and decrease sector weightings.

The style enables it to be used for both independent learning (i.e., online courses) and as a standard textbook. Examples and exercises will be offered online so that they are capable of being updated without requiring the entire book to be republished.

At the beginning of each chapter, we highlight the objectives and key concepts relevant to that chapter.

At the end of each chapter, there are a series of quiz questions to test understanding. The answers are shown in the Appendix. There are also a series of questions for discussion in the classroom, as well as suggested topics for dissertations.

We have also included as an online resource the spreadsheets used for the chapters, which will allow students to input their own assumptions and understand more fully the concepts explained in each chapter.

Global real estate markets

1 Global real estate in context

Objectives

At the end of this chapter, you will understand the:

- Background to real estate as an institutional asset class.
- Reasons for this book.
- Relative size of the real estate market.
- Quadrant Model.
- Structure of this book.

1.1 Background

Real estate has a long history as an institutional asset class. Some of the major holdings of the largest landowners in the UK, the colleges of Oxford and Cambridge University, date back to the 16th century, and still form a significant part of their investment portfolios, with the income generated from them funding their day-to-day operations. More recently, we have seen the growth of institutional real estate allocations during the 1960s as pension funds sought the long-term leases available on the office buildings developed at the time to provide income to match their liabilities within Defined Benefit Schemes. This continued with the development of multi-tenanted regional shopping centres, which provided both scales, and until the full impact of the advent of E-commerce was felt, certainty and stability of income.

Real estate was thus seen as a provider of long-term income and stable cash flows, with the land component providing a hedge against inflation, a low volatility and a correlation to equities and bonds. In many ways, therefore, it was the perfect complement to the typical 60/40 equity-bond portfolio. Its role was further helped by the work of David Swensen, the CIO at Yale University, who developed the popular and hugely successful endowment model, which *inter alia*, popularised exposure to illiquid, "alternative" assets (i.e., not equities or bonds). Thus, the main criticism of real estate, that it was illiquid, was turned into an advantage. Managers were not seen as holding lumpy assets at the wrong valuation, but instead were "harvesting the illiquidity" premium. In addition, at a time when the volatility of liquid assets was increasing the low volatility and therefore the higher risk-adjusted returns of an illiquid asset with smoothed returns due to the appraisal nature of valuations, real estate enhanced its attraction further to investors.

Fast forward to the current day and what has changed? We think four key trends have led to significant structural changes in the role of real estate in investment allocations and will continue to do so.

DOI: 10.1201/9781003298564-2

1.1.1   Globalisation

Real estate used to be seen as a local business. A good example (and there are plenty) of this was in Ireland before the global financial crisis of 2007. Irish developers who were (at times 100%+) financed by Irish banks, sold assets to Irish funds, and the funds were sold to Irish investors. This led to a perfect storm in terms of systemic risk, and when the inevitable fall in values came, the repercussions were felt across the entire Irish financial system. To this day, the actions of the Central Bank of Ireland are focused on avoiding a repetition of this.

Whilst there had been selected overseas investors before, mainly from markets enjoying a real estate and lending boom, such as the Japanese and New Zealand investors in the 1980s, Nordic investors in the 1990s, these forays were normally short-lived and ended badly. As a result, this type of investor became known as a "leveraged tourist", and their entry was normally the signal for domestic investors to exit the market.

It was not really until the GFC that capital markets became truly global on an ongoing basis. Part of the reason for this lay in the globalisation of monetary policy (quantitative easing) to restore confidence in the financial system. As a result, global investors started to compare overseas investment yields with their own. This resulted in, for example, the sale of several "trophy" assets in London and elsewhere, with investors selling at well below historically low yields but above the yields of the buyer's local market (in a number of high-profile cases, this was Hong Kong). This globalisation has remained and indeed one of the key factors in assessing a market is the depth of liquidity, that is, the level of overseas and particularly first-time buyers. Therefore, we take a global view in this book. Certainly, all securitised assets (i.e., real estate investment trusts and bonds/commercial mortgage-backed securities (CMBS)) will be impacted by global events as will all real estate assets above a certain size or in a major city location. Globalisation is therefore here to stay.

1.1.2   Emergence of real estate debt as an asset class

Whilst the modern REIT era started in the early 1990s and the boom in private real estate funds started in the early 2000s, it is noteworthy that post-GFC, we have seen the emergence of real estate debt as a separate and growing asset class. Part of the reason for this is regulatory, with institutions forced to provide more capital for an asset than a loan. But part is also market opportunity. As banks have legacy issues and constraints on loan-to-value (LTV), the opportunity exists for non-traditional lenders to enter the markets. This could be in the form of mezzanine funds, whole loan funds or insurance funds. We believe this trend will continue.

1.1.3   Growth of "alternative" sectors

The lasting impact of the pandemic in 2020 and 2021 was to accelerate the structural changes that were occurring in the direct property market. The decline in the appeal of two of the core sectors, offices, and retail, due to the emergence of "working-from-home" (WFH) and online rather than in-shop retailing. Arguments will continue about the nuances of the impacts of these changes, but one point is clear: Aggregate demand for offices and retail space is unlikely to increase. Obviously, certain product types, locations, industries, and sub-markets will prosper, and existing assets can be repurposed into alternative uses (subject to planning), but overall, the total demand for existing offices and retail in current use is unlikely to increase in any material way. This has already led to a rotation of capital allocated to those assets into industrial/logistics and, more importantly, the "alternative" sectors. These sectors include residential, healthcare, data centres, cell towers, life sciences, student accommodation and self-storage. We believe that capital will continue to be drawn to these sectors and that the future growth of the sector will come from demand for these uses. These are covered in more detail in Chapter 3. As a result, and because of the divergent demand factors driving these sub-sectors, performance for the sector overall will be difficult to determine. This divergence and huge sector-wide dispersion around an industry average started in 2017, and we believe it will continue.

1.1.4   *Environmental, social and governance (ESG) perspective*

Finally, it is impossible to escape the impact that ESG issues are having on allocations. Rather than a linear gradation, the impact has been effectively binary. Either funds or REIT schemes demonstrate sufficient resilience in this area, or they do not. If they do, then capital (both debt and equity) will be allocated; if not, then they become "stranded". The biggest challenge the industry currently faces is how to effectively price in the cost of retrofitting assets to the new institutional standard.

1.2   Reason for this book

There are currently no comprehensive textbooks available that focus on both the academic and practical applications of real estate capital markets (i.e., not valuation or real estate economics) in global real estate. Our approach is to combine a structured, rigorous understanding of the theory and academic evidence behind the main concepts with practical examples and applications, utilising the authors' own extensive experience.

The book is designed for real estate students, both under and post-graduates, and also will be used as the main textbook for global finance courses, for Executive Education short courses and as a primer for new entrants to the sector. It is aimed at the following groups who are not in full-time tertiary education:

1 Property professionals working for a listed company wanting to understand the relationship between their underlying business and the stock market valuation
2 Real estate private equity teams looking to understand the valuation disconnect between public and private markets and arbitrage the Parallel Asset Pricing model
3 Equity/multi-asset/property analysts/fund managers who need to understand the specific characteristics of real estate vs. the other ten equity sectors and understand when to increase and decrease sector weightings.

The style will enable it to be used for both independent learning (i.e., online courses) and as a standard textbook. Examples and exercises will be offered online so that they are capable of being updated without requiring the entire book to be republished.

The book enables readers to understand, for example:

• Why share prices of real estate companies can differ dramatically from the underlying value of the assets
• The differing investment objectives of different categories of investors and how this influences share prices and corporate funding decisions

- How sell-side analysts make their recommendations
- How buy-side analysts decide which sectors, funds, and stocks to allocate capital to
- How ESG considerations are relevant to capital market pricing. We address these issues in each relevant chapter rather than in a separate ESG chapter.

This book unravels the complex mechanisms involved in global real estate capital markets, enabling the reader to understand how they have grown and evolved, how they function, what determines market pricing, and how the public and private debt and equity markets are linked to each other.

1.3   Global real estate capital markets in context

1.3.1   *Relative to other assets*

It is important to understand the significance of the global real estate capital market within the framework of investible assets available to institutional investors. How, for example, does the size of the commercial real estate market compare with equities, bonds, and other investments?

Figure 1.1 shows that the largest investible market is the bond market, at a value of over US$100 trillion. This is roughly equivalent to the size of oil reserves globally and 25% larger than the global equity market. As an asset class (including owner-occupied and non-investible assets), real estate is larger, at around US$280 trillion, although it should be noted that the most significant part of that is residential, where the vast proportion is held for occupational use rather than investment use. Commercial real estate is still extremely meaningful, though at around a third of the size of the bond market.

Global real estate universe in comparison

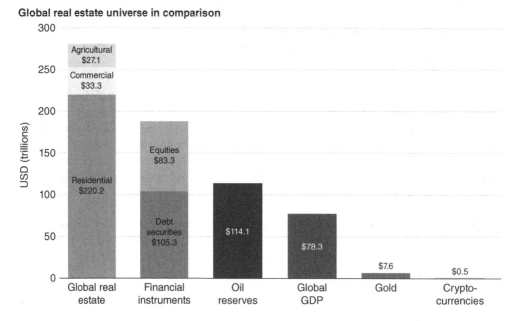

Figure 1.1 Global real estate size relative to other assets.

Source: Savills World Research.

EPRA estimate that the total value of commercial real estate at the end of 2022 was around US$34 trn. They provide a useful breakdown so that for each country, you can see the size of the economy (as measured by GDP), the size of the commercial real estate market, and the size of the listed sector. Figure 1.2 provides a snapshot of Europe as an example.

1.4 The quadrant model

The real estate quadrant approach for categorising the real estate investment universe is now over 20 years old. Institutional investors have four primary investment conduits through which they can gain exposure to commercial real estate: public and private holding structures across debt and equity investments.

These four investment options are often referred to as the "quadrants" and can be accessed through a range of holding structures, including direct ownership and management or indirect exposure through funds (such as limited partnerships) and other investment vehicles. Investors are often invested in more than one of these quadrants, whether this be in discrete real estate programmes or when considering their portfolios in a multi-asset context. For example, REITs may form a component of an investor's dedicated real estate allocation or be accessed through their public markets investments if this is not the case. However, evolving institutional real estate investment strategies, driven by both market and regulatory forces, are leading real estate investors to re-examine the broadening of blended strategies, with active allocations taking place across multiple quadrants.

The quadrant investment model was first illustrated by **Hudson-Wilson (2001)** and further developed in subsequent publications (**Hudson-Wilson, Fabozzi, and Gordon 2003**; **Hudson-Wilson et al. 2005**). It is worth noting that in the original paper, the idea was that the universe should be expanded from the traditional realms of private equity and private debt to include the "new" public equity and public debt market exposures. The real estate quadrant universe has developed considerably over the past 20 years and has become increasingly accessible for investors seeking exposure to the asset class. The development and performance of the quadrants during the recent market cycle warrants a re-examination of their risk and return characteristics.

We can look at the four quadrants as comprising the following asset classes and constituting the following relative contributions to the overall market (Figure 1.3).

Over the past 20 years, the relative size of the component quadrants for the US has changed as follows (2001 figures in parentheses): Private debt remains the largest, at 47% (46%). Public debt at 25% (12%) has shown the most significant relative growth. Private equity has declined to 15% (33%) and public equity has risen to 13% (9%). Overall, private markets remain the most significant, at 62% versus 38% for public markets. This compares with the 79%:21% split in 2001.

1.5 Structure of this book

Each chapter of this book has a series of quiz questions, as well as topics for discussion and further research/dissertation. Excel tables and worked examples are available online.

The book is divided into five areas, covering:

Global Real Estate
Direct Markets

Size of the total commercial real estate market - Developed Markets

E P R A

	Dec-21 GDP per Capita ($)	Dec-21 GDP ($ Bln.)	Dec-21 Commercial Real Estate ($ Bln.)	Sep-22 Total Listed Real Estate ($ Bln.)	Sep-22 Number of Companies (#)	Sep-22 REITs Market Cap ($ Bln.)	Sep-22 Of which REITs (#)	Sep-22 Non-REITs Market Cap ($ Bln.)	Sep-22 Of which Non-REITs (#)	Sep-22 Stock Market Size ($ Bln.)	Sep-22 Listed RE/Stock Market %	Sep-22 Listed RE/Total CRE %
Austria	53,793.37	481.21	205.56	6.91	7	-	-	6.91	7	115.29	6.00%	3.36%
Belgium	50,412.71	581.85	246.72	24.71	29	17.97	17	6.75	12	296.81	8.33%	10.02%
Denmark	67,919.59	396.67	166.81	2.31	8	-	-	2.31	8	533.71	0.43%	1.38%
Finland	53,522.57	296.02	126.28	4.50	6	0.03	1	4.47	5	225.86	1.99%	3.56%
France	45,028.27	2,940.43	1,254.70	38.94	47	34.56	28	4.38	19	2,322.49	1.68%	3.10%
Germany	50,787.86	4,230.17	1,809.33	65.82	53	2.41	6	63.42	47	1,630.27	4.04%	3.64%
Ireland	102,394.02	516.25	203.63	0.60	1	0.60	1	-	-	80.10	0.75%	0.29%
Italy	35,584.88	2,120.23	910.76	0.78	8	0.38	2	0.41	6	459.27	0.17%	0.09%
Luxembourg	131,301.60	83.77	34.55	-	-	-	-	-	-	11.29	-	-
Netherlands*	57,714.88	1,007.56	427.71	8.29	7	8.21	5	0.08	2	760.01	1.09%	1.94%
Norway	82,244.23	445.51	185.29	4.20	8	-	-	4.20	8	336.34	1.25%	2.27%
Poland	17,945.75	679.07	234.96	3.84	35	-	-	3.84	35	107.01	3.58%	1.63%
Portugal	24,457.14	251.71	99.87	0.09	3	0.06	2	0.03	1	72.99	0.12%	0.09%
Spain	30,536.86	1,439.96	615.12	24.73	89	20.22	75	4.51	14	510.31	4.85%	4.02%
Sweden	58,639.19	622.37	258.11	45.61	48	-	-	45.61	48	763.67	5.97%	17.67%
Switzerland	93,515.48	810.83	346.50	52.36	48	-	-	52.36	40	1,666.85	3.14%	15.11%
United Kingdom	46,200.26	3,108.42	1,636.33	64.27	80	58.36	54	5.91	26	2,474.44	2.60%	3.93%
Total Europe		20,012.01	8,762.25	347.97	469	142.78	191	205.19	278	12,366.72	2.81%	3.97%

Figure 1.2 Size of the total commercial real estate market.

Source: EPRA.

	Public	**Private**
Debt	Fannie/Freddie, CMBS, CMO, Mortgage REITs, REIT Corp Bonds	Senior/Whole Loans
	Size: US$1,376bn	*Size: US$2,541bn*
Equity	REITs	Directly held assets, LPs, LLCs
	Size: US$697bn	*Size: US$1,376bn*

Figure 1.3 The real estate quadrant model.

Source: Prepared by the authors from Pension Real Estate Association data, August 2020.

Private Markets
Public Markets
Applications.

Within these five main headings, the breakdown of topics and key points covered in each chapter is as follows:

1.5.1 Global real estate

1.5.1.1 Global real estate in context

The background to real estate as an institutional asset class, the gap in the market that this book fills, the relative size of the real estate market, and the real estate quadrant model.

1.5.1.2 Understanding different investors' objectives and strategies

This chapter looks at the different investment objectives, time horizons and strategies of every type of potential shareholder.

1.5.2 Direct markets

1.5.2.1 Direct real estate capital markets

A foundational chapter describing commercial real estate as an asset class, covering both traditional and 'other' property types–operational aspects and key considerations. The market universe is also described.

1.5.2.2 Analysing a direct real estate investment

The basics of lease structures, net operating income (NOI) projections and end capitalisation are presented with reference to prospective performance measures such as IRR. The impact of utilising financial leverage is also presented and the implications of changing assumptions upon underwritten returns.

1.5.2.3 Valuing direct real estate assets

Key concepts and definitions (e.g., the various yield/cap rate definitions) are presented, along with the main methodologies employed across the major regions. How the valuations of non-traditional sectors are arrived at is also addressed.

1.5.3   Private markets

1.5.3.1   Private real estate fund structures

What are the key types of private real estate structures and their strategies? This chapter outlines the key differences between open-ended funds, close-ended funds, and Perpetual Life and Finite Life Funds across risk-return profiles. Potential agency issues are covered.

1.5.3.2   Valuing private real estate funds

What are the key accounting methodologies employed and how does this impact measured performance? The impact of market liquidity on certain structures and the developing secondary market are also covered.

1.5.3.3   Private debt structures

This chapter looks at the importance of security, covenants, and terms in private debt structures. Potential agency issues are covered.

1.5.4   Public markets

1.5.4.1   Public market structures

This chapter introduces the reader to the structure of the market and provides more depth on the key market participants and industry bodies. Data and benchmarking are presented.

1.5.4.2   Analysing public real estate equity

To analyse any listed company globally, only five key metrics are needed. This chapter outlines what they are, what they represent, and the key issues to address in each of them. Practical examples from companies' investor presentations are used.

1.5.4.3   Analysing public real estate debt

In this chapter, we deal with the real estate quadrant that represents public debt. At its broadest definition, it covers Fannie Mae and Freddie Mac (these are specific to the US, representing federally backed home mortgage companies), commercial mortgage-backed securities ("CMBS"), residential mortgage-backed securities ("RMBS"), collateralised mortgage obligations ("CMO"), mortgage REITs, and REIT corporate bonds. In this chapter, we focus exclusively on REIT and REOC corporate bonds. We will look at the key characteristics and terms of a bond, how investors look at the universe of opportunities in REIT bond markets, key trends in bond issuance globally and, finally, at the broad methodologies of credit rating agencies.

1.5.4.4   Valuing a public real estate company

How do you value a listed real estate company? This chapter outlines the three basic approaches that are used in practice, allowing the reader to value any listed real estate company globally.

1.5.4.5   Setting target prices and recommendations

For listed companies, all sell-side analysts set a target price and Recommendation. What is a target price, how can it be calculated, and what does it mean?

1.5.4.6   Understanding the operations of a public real estate company

This concluding chapter for Section 2 looks at how to understand what a share price represents, and how and why it can diverge from the underlying value of the assets of a company at different stages of the cycle.

1.5.5   Applications

1.5.5.1   Bringing it all together

There are several key topics which are not dealt with in textbooks about the practical implications of implementing a public real estate strategy. These include smart beta, screening and blending.

2 Understanding different shareholders' objectives

Objectives

At the end of this chapter, you will understand how to classify and understand the motivations of different investment institutions according to their:

- Size, measured in terms of assets under management ("AuM").
- Investment time horizon.
- Return and performance measurement criteria.
- Use of listed real estate.

Key concepts

- Different groups of institutional investors have different approaches to asset allocation depending on their organisational structure and mandate.
- The attractions of listed real estate are for some institutions time variant, while for others they form a permanent allocation.
- Understanding and identifying different shareholder objectives is key to growing a company's size, as the shareholder base needs to evolve to match equity capital-raising objectives.

2.1 Size

The first criterion we can use to classify different investment institutions is their size, which we typically look at in terms of the AuM of that institution.

To put this into context, at the end of 2022, according to IPE Real Assets the fund managers (i.e., third-party managers not including sovereign wealth funds or pension funds) with the largest real estate AuMs were as follows (Figure 2.1).

This shows the dominance of the largest two institutions, with the largest, Blackstone, having twice the AuM of the fourth largest asset manager. These figures relate to global real estate assets held directly in JVs, separate account mandates and via collective vehicles/funds.

It is important to understand that breaking down the assets held by geography (US, Europe, Asia), sector (office, logistics) and style (Core, Core-Plus, Value Add and Opportunistic) will highlight the specialisations of the different asset managers. For example, AXA IM has the largest real estate exposure in Europe, and ESR, the largest AuM in Asia. Predictably Blackstone is the largest in the US. In terms of sectors, Brookfield is the largest office owner globally amongst the managers, although in Europe,

DOI: 10.1201/9781003298564-3

Company	AuM (€m)
Blackstone	305,070
Brookfield Asset Management	236,206
MetLife Investment Management	182,573
PGIM Real Estate	154,245
Nuveen	149,083
ESR Group	132,231
JP Morgan Asset Management	128,368
CBRE Investment Management	128,300
AXA IM Alts	116,104
Starwood Capital Group	110,768

Figure 2.1 Global real estate AuM.

Source: IPE Real Assets.

German managers (Deka and Allianz) are the most significant. Whilst Prologis is the largest logistics operator in the Americas, it is GLP which is the largest globally due to its holdings in Asia. We can see therefore that it is important to drill down into an asset manager's geographic, sector and style preferences to determine their potential asset allocation appetite.

In addition to standard classifications, there are two sub-sectors worth highlighting.

The first of these is fund of funds/multi-manager AuMs. These are groups who provide advisory services (discretionary and non-discretionary) to institutional clients and allocate capital to different (predominantly) externally managed funds. They play an important part in the capital allocation process. The largest of these managers by size at end 2022 was as follows (Figure 2.2):

Company	AuM (€)
CBRE Investment Management	38,700
UBS Asset Management	31,714
Stepstone Group Real Estate	10,439
LaSalle Investment Management	7,550
Ivanhoe Cambridge	5,888
Schroders Capital	5,214
Bentall GreenOak	3,105
Credit Suisse Asset Management	2,728
IGIS Asset Management	2,629
HIH Real Estate Investment	2,111

Figure 2.2 Fund of funds/multi-manager AuM.

Source: IPE Real Assets.

Company	AuM (€)
Cohen & Steers Capital Management	47,573
CapitaLand	39,748
ESR Group	33,153
Starwood Capital Group	27,597
Mapletree Investments	26,751
APG Real Estate	25,966
UBS Asset Management	21,541
Principal Real Estate Investors	20,899
Invesco Real Estate	15,054
CentreSquare Investment Management	10,830

Figure 2.3 Listed real estate/REITs AuM.

Source: IPE Real Assets.

The final group is managers of listed real estate /REITs. It is important to understand that there are two categories within this group. The first are the dedicated REIT portfolio managers, such as Cohen and Steers. They manage funds which hold listed real estate. The second group includes names such as CapitaLand. They own asset management platforms that manage publicly listed REITs, and they may also take direct stakes in listed real estate companies (Figure 2.3).

Clearly, these numbers vary from year to year, but the tables above provide an insight into who the main participants in the market are and their scale.

We now turn to classification. How can we break down the different participants by common groupings based on their overall size and investment preferences? We believe that there are six different types of investors to consider, and we use these groupings throughout this chapter. They are:

1 Private individuals
2 Wealth managers
3 Ultra-high net-worth wealth managers
4 Institutional asset managers
5 Pension funds
6 Sovereign wealth funds

Family offices have not been included in this list, even though they are important as it is very difficult to make general assumptions about them. The smallest family offices fall more under the category of private individuals whilst the largest can be more accurately grouped in terms of characteristics (but not size) under sovereign wealth funds. Also, it is important to note that these size groupings are based on individual, not aggregate, size. In other words, most sovereign wealth funds are larger than most pension funds and most wealth managers will be larger than individual asset owners. However, in aggregate, institutional asset managers may have a larger total AuM than pension funds in total. In addition, it is worth noting that one must be wary of double counting. A sovereign wealth fund may, for example, own US$100bn of assets, of which US$60bn is managed by institutional asset managers. These managers would also show that US$60bn under their AuM.

We now look at each of the six groups (shown in ascending order) and highlight key points about their asset allocation.

2.1.1   Private individuals

- Individuals typically invest in REITs (or unlisted funds) having done their own research or on the advice of a broker.
- Individuals provide liquidity into the market, as typically they are dealing in a small lot size with "frictionless" trades (i.e., their trades do not alter prices).
- Individuals' trade executions can normally be accommodated in the market size quote.
- Research is typically from tipsheets/paid-for research providers/stockbrokers.

2.1.2   Wealth managers (discretionary aggregated private client accounts)

- This sector has shown dramatic growth in the last ten years due to sector consolidation. As an example, Investec Wealth Management merged with Rathbones to create a £100 billion AuM institution.
- The largest are as powerful as institutional asset managers and are often staffed by ex-institutional fund managers.
- They have been known to hold >10% positions and be the largest group of shareholders in smaller IPOs.
- Typically, they hold individual stocks/funds for specific asset exposure (e.g., life science REIT, digital realty, Prologis) or diversified vehicles as a tactical/core exposure to the sector.
- In the latter case, they often prefer less volatile (smaller) stocks/funds to replicate direct exposure, minimising equity market influences on pricing.

2.1.3   Ultra-high net-worth wealth managers (Swiss/US banks) and family offices

- Typically, those with >US$1m of investible assets are regarded as high net worth (HNW) individuals and those with >US$30m of investable are known as ultra-high net worth (UHNW) individuals.
- Well-known examples of institutions who deal with both groups are UBS and Citibank.
- The higher the investible amount, the more customised the investment solution provided.
- Firms often sponsor or create structured products where listed real estate has a role.
- Clients' requirements are normally very specific.
- Often the products used are leveraged or have variable leverage.

2.1.4   Institutional asset managers

- Well-known examples include Blackrock, Fidelity, and Schroders.
- The main distinction in their product range is between active and passive products.
- Most offer a suite of funds to investors including general equity and specialist real estate products.
- In addition to funds (pooled vehicles), separate account mandates can also be offered.
- Funds can be absolute or relative index benchmarked.

Name	Country	AuM US $bn
China Investment Corp (CIC)	China	1,351
Norges Bank Investment	Norway	1,145
Abu Dhabi Investment Authority (ADIA)	UAE	993
State Admin of Foreign Exchanges	China	980
Kuwait Investment Authority	Kuwait	769
Got of Singapore Investment Corporation (GIC)	Singapore	690
Public Investment Fund	Saudi Arabia	620
Hong Kong Monetary Authority	Hong Kong	500
National Council for Social Security	China	474
Qatar Investment Authority	UAE	450

Figure 2.4 Leading global sovereign wealth funds 2022, by AuM.

Source: Statista.

2.1.5   *Pension funds*

- Examples: PGGM, APG
- The main distinction is between defined benefit (DB) and defined contribution (DC) schemes.
- A significant consolidation amongst managers is occurring globally, with a heavy focus on administrative costs and net returns.
- A good example of this is the Legal and General DC mandate for a NEST for blended real estate products.

2.1.6   *Sovereign wealth funds*

Figure 2.4 shows the largest sovereign wealth funds at the end of 2022 by AuM. Note that the AuM shown is total assets, not just real estate. As can be seen, the largest controls US $1.4 trillion of assets.

- These funds have been hugely influential over the last ten years.
- They can invest across the capital stack, for example, GIC owning equity and debt, instruments of British Land, as well as having a JV with them. This is similarly the case with Norges and Prologis.
- Potentially this is the largest pool of investment capital; therefore, allocations are limited to larger companies.

We have identified the diverse types of investor groups, understood their basic rationale, and differentiated features, and shown examples of each type.

2.2   Differing time horizons

We now turn to the different time horizons and objectives of the various investor groups identified. For each category, we show examples of their requirements and the potential uses of listed real estate.

2.2.1 Private individuals

- Income-based REITs and REIT funds are popular with investors reliant upon passive income sources (e.g., pensioners).
- Monthly or quarterly payouts may be especially attractive.
- The time horizon for these income investors is typically long-term.
- It is dependent upon the age of the individual.
- Broadly speaking, the younger the investor, the greater the risk appetite = the greater the proportion of total return to be in capital rather than income, and the shorter the investment time horizon.
- Prima facie, this suggests REITs with their minimum payout should attract an "elderly" investor; however, this ignores the impact of leverage on capital returns.

2.2.2 Wealth managers

- Typically, their standing depends on two factors, the market cycle, and the stage of a company's evolution:
- The longevity of the cycle if the investment is for tactical reasons.
- Company growth, both in terms of fundamentals and capital raises.
- Wealth managers typically prefer lower volatility and more infrequent cash calls.
- The more a company raises fresh equity and becomes part of equity indices, the more the wealth manager's stake becomes diluted.

2.2.3 UHNW managers

- Time horizons can be anything - from very short-term tactical (e.g., a Brexit basket) to longer-term structural (investment in residential across Europe).
- A number of products have been developed that allow the investor to select the leverage applicable to them.
- The investment can be in anything from a quant-driven fund to an actively managed (higher fees) separate account.

2.2.4 Institutional asset managers

- Average holding periods for institutional asset managers are estimated at 3–5 years. However, there are many exceptions.
- Most absolute return allocations will be given cap and collar review prices to see if the strategy is working. Holding periods can range from days to several years.
- For benchmark weightings, a Core holding of the largest stocks in the index will be broadly permanent, although it is subject to weighting rebalancing on a regular basis to try and outperform the index.

2.2.5 Pension funds

- Typically, there are longer-term time horizons.
- Holdings tend to be significant, so a large amount of due diligence is undertaken (sunk cost).
- In addition, due to the size of the holding (3%–20% is not uncommon), it is impossible to trade out quickly without affecting the price. The normal option is a block sale or book-building exercise.

2.2.6   *Sovereign wealth funds*

- Potentially, very long-term time horizons are involved, partly due to the same reasons as those for pension funds.
- Note that they may exit one, but not all parts of the capital stack.
- As mentioned, Norges holds Prologis equity has a JV portfolio with them and possibly own their bonds. They may increase or decrease their weighting in any of these to reweight their overall exposure to the company.

We have now looked at the differing time horizons of the different investor groups and understood how their size and structure can affect both their portfolio decisions and investment universe.

2.3   Allocation, return and performance criteria

Understanding the different return criteria of investors is key to understanding their allocation policies. Below, we highlight the key aspects of each group:

2.3.1   *Private individuals and wealth managers*

Capital preservation is often key. Therefore an 80/20 return split (income/capital) is often desired.

For most individual clients, there is little need for additional residential exposure as they typically own equity in at least one residential asset.

2.3.2   *Ultra-high net-worth wealth managers (Swiss/US banks) and family offices*

They provide a multitude of investment strategies, utilising both strategic and tactical exposure across the capital stack. Products offered can include one or all of listed, direct, unlisted, debt, and derivatives.

2.3.2.1   *Institutional asset managers*

The performance of most funds versus the benchmark (absolute or relative) is key to gaining AuM and retaining clients. They typically provide a detailed explanation of the components of the expected return as well as expected volatility, liquidity and tracking error.

2.3.2.2   *Pension funds*

For DB schemes, the return criteria are normally asset/liability matching. Therefore, a high-income component to total returns with capital preservation in real terms (i.e., after adjusting for inflation) is a requirement. For DC schemes, the purpose is normally to add liquidity to existing real estate exposure.

2.3.2.3   *Larger pension funds (such as APG) and sovereign wealth funds*

These funds offer a strategic deployment across the capital stack, including private debt and underwriting. They may have board representation and up to a 20% stake in individual companies.

2.4 Sample applications of listed real estate

Finally, we look at examples of how listed real estate has been used by different investor groups.

2.4.1 Non-benchmarked funds

Non-benchmark-constrained funds seek to provide what is typically described as "a mixture of income and capital growth". A good example of a non-benchmark-constrained fund is the *M&G Global Real Estate Securities Fund,* set up in 2008. As a result of this investment process, the fund weightings can be quite different from a benchmarked fund. Figure 2.5 illustrates the portfolio holdings of the M&G Fund compared to a leading benchmark fund, in this case, the *ING Global Real Estate Fund,* which is benchmarked to the S&P Developed Property Index. As can be seen, there are notable differences in exposure to, for example, the US, Japan and Canada, as well as niche companies such as Shaftesbury, which reflect, *inter alia*, the different sizes of the fund (the ING has assets of US$4,500 m while the M&G Fund is about US$100 m), as well as the choice of benchmark (both funds have four-star ratings from Morningstar).

The best way of understanding the investment process is to look at the formal definition in the prospectus, which states the following:

Top 10 holdings %			
ING Fund		**M&G Fund**	
Simon Property	4.7	Simon Property	6.2
Mitsubishi Estate	4.0	Westfield Group	4.1
Mitsui Fudosan	3.3	Link REIT	4.1
Westfield Group	3.1	Unibail-Rodamco	2.7
Boston Properties	2.8	Avalonbay Communi	2.7
Unibail - Rodamco	2.7	CBL & Associates	2.6
Cheung Kong	2.4	Summarecon Agung	2.5
Host Hotels	2.3	Shaftesbury	2.5
Macerich	2.2	Boardwalk	2.4
Westfield Retail Trust	2.0	Land Securities	2.4

Top Country Weightings %			
ING Fund		**M&G Fund**	
United States	43.6	United States	36.6
Japan	14.2	Hong Kong	11.1
Hong Kong	12.3	Canada	8.6
Australia	10.1	Australia	8.0
United Kingdom	5.3	UK	7.0
France	5.1	Japan	6.3
Singapore	5.1	France	2.7
Canada	1.7	Indonesia	2.5

Figure 2.5 Portfolio allocations, benchmarked and non-benchmarked.

Source: Fund fact sheets.

The M&G Global Real Estate Securities Fund invests in real estate investment trusts (REITs) as well as other types of property companies to optimise long-term total returns. The fund manager aims to exploit market pricing opportunities through the selection of global property stocks that are undervalued by investors, with overall portfolio construction also informed by top-down regional and sector views. The overlay of the fund takes into account real estate dynamics and risk assessment. A high-conviction investment process is utilised to select stocks, with a focus on structural shifts in global property markets through listed, rather than direct, property vehicles. While remaining risk-aware, there is a pragmatic approach to stock selection. Attractively priced stocks are screened on an absolute basis, rather than on relative value, in the belief that real estate prices are driven by the NAV of a company.

2.4.2   Specialist income funds

While both benchmarked and non-benchmarked funds seek to provide a mixture of income and capital growth, the decline in risk-free rates globally has led to a rise in demand for income-focused listed real estate products. Two specific strategies seek to deliver this: first, using a portfolio of higher-yielding REITs; second, through writing call options on the underlying portfolio and taking the premium received, distributing it as a dividend to unit holders.

Good examples of the former strategy are the *CBRE Clarion Global Real Estate Income Fund*, the *Cohen & Steers Realty Income Fund*, and, on a regional basis, the *B&I Pan-Asian Total Return Real Estate Securities Fund*. A good example of the second strategy is the *Schroder Global Property Income Maximiser Fund.*

The key issue to remember with these enhanced income funds is that investors accept a cap on future performance above a certain level (by writing a call option) and taking that as income. It is therefore suited to certain stages of the cycle when capital growth is limited, and income is required.

2.4.3   Long/short, 130/30 and hedge funds

All of the above fund types run long-only strategies. In other words, if the fund manager has a bearish view of the market, the most defensive action he can take is to increase the level of cash to the maximum allowed under the fund's prospectus (normally 10%). During the mid-2000s, a new type of fund emerged which allowed long-only managers to have limited short positions. These were known as 130/30 funds. In essence, they allowed a fund manager to run a maximum 130% long exposure which would be offset by a maximum 30% short position. If, for example, a fund had assets of £100 m and the manager was positive on the UK but negative on Europe, then instead of a £100 m UK exposure, they would be allowed £130 m, enhancing returns assuming the position proved to be profitable, offset by a short position of £30 m on European stocks, leaving a net £100 m exposure. Several fund managers launched such funds. They did not prove overwhelmingly popular initially but are now becoming more popular.

The well-established team at Thames River has run a similar UCITs 111-compliant style long-short fund and is currently launching an *F&C Real Estate Equity Long/Short Fund* which aims to generate a return of 8–10% by investing in a portfolio of pan-European real estate securities. *Cohen & Steers* also runs a *Global Real Estate Long/Short Fund*, and it is worth examining its investment philosophy.

Long positions focus on property companies trading at a discount to their net asset values, with strong balance sheets, proven management teams and solid business models. Short positions are used as opportunistic investments based on specific catalysts, or as hedging instruments designed to mitigate risks related to the broad equity market, foreign exchange rates, interest rates and various country- and company-specific factors.

2.4.4 Using listed real estate as a proxy for direct real estate

In the US, a strategy of using listed real estate as a proxy for direct real estate has been extremely popular with US state pension funds, which have awarded domestic and global REIT mandates to fulfil their requirement for a real estate exposure, whilst in Europe, PGGM and APG have been keen proponents of using the listed sector as a way of delivering direct real estate returns. An example is the China National Council for Social Security Funds mandate, awarded to AMP. China's National Council for Social Security Fund (NCSSF) is a €111 bn sovereign fund. Although the fund was established in 2001, the NCSSF was prohibited from investing overseas until 2006. Having appointed its first global equities, fixed income, and regional mandates, it is now targeting global REITs as a liquid, tax-efficient proxy for the global real estate market. Anthony Faso, international business director at AMP Capital, said:

The NCSSF has reached out to managers to develop its expertise in global investment. They want to learn about pensions and to evolve towards the international market.

The mandate seeks to outperform the EPRA benchmark, with 32% allocated to Asia, 14% to Europe and the balance to North American markets.

Dutch pension fund BPF Bow increased its allocation to listed real estate as part of a plan to invest in both public and private real estate in a fully integrated way. It invests as much as 8% of its real estate capital in listed investments as part of a portfolio managed by a single, integrated team at asset manager Bouwinvest Real Estate Investment Management.

The fully integrated listed strategy enabled Bouwinvest to take advantage of market cycles and arbitrage opportunities with the non-listed sector. Another main objective was to "fill in holes", where Bouwinvest is unable to find the right non-listed product, whether this is a real estate fund, a club deal, or a joint venture. A recent example of the execution of this strategy was an investment in five REITs to gain exposure to regional malls in the US, given that many of the best assets in the sector are held by listed REITs.

2.4.5 Using listed real estate for platform investing by asset managers

The concept of platform investing is well-known to real estate private equity managers, but it is increasingly being used in the listed sector. There are two reasons for this: firstly, a belief that specialist management teams and niche sectors deliver superior returns over the long term; secondly, the current limited equity funding available for smaller listed companies means that platform investors can fill the gap, providing a critical lifeline of support and expansion in the same way that mezzanine debt funds fill the gap left by the reduction in senior debt.

In a market where there is often a focus on liquidity, companies such as *Forum Partners* have helped to provide restructuring capital and market expertise to small and mid-size real estate companies that have been traditionally underserved by the capital markets. In return for providing equity or mezzanine finance, equating to say 10–25% of the company's capital structure, they take a seat on the board and help the company to expand. They have successfully executed this strategy globally, with over 70 investments made in 17 countries in Europe and Asia. Examples of their investments in Europe are New River Retail in the UK and Zueblin in Switzerland.

Quiz questions

1 Name five of the largest real estate managers globally.
2 What are fund of funds/multi-managers?
3 What are the two types of listed real estate manager?
4 What are the six different classifications of investor?
5 Which countries have the largest sovereign wealth funds?

Discussion questions

• What determines the allocation policy of an institutional investor?
• How would the strategy of a private investor differ from that of a UHNW investor?
• How do the DB and DC pension fund strategies differ?

Research/Dissertation topics

Investing across the capital stack (i.e., public equity, public debt, private debt, private equity) should allow SWFs and others to optimise performance across the cycle.

3 Direct real estate capital markets

Overview

In this chapter, we provide an overview of the structure of the direct real estate market from the perspective of a global institutional investor. In later chapters, we will examine the structure of the public and private asset owners of these direct assets. We define the key terms and concepts which are unique to real estate and specify the different occupational categories of investible real estate. Finally, we define market size and the roles of the key participants.

Objectives

At the end of this chapter, you will understand:

- The key ownership types, lease structures and occupational markets involved in the institutional commercial real estate market.
- How to define the market size globally.
- The key participants and motivations of the sell-side and buy-side.
- How new investible stock is created through developments sales, and leasebacks.

Key concepts

- Ownership and occupational lease structures.
- Definitions of market size and comparison with other capital markets.
- Interaction of market participants and different motivations of buy and sell-side.
- Growth of the market through new investible stock.

3.1 Direct real estate

3.1.1 *Definition*

The Cambridge Dictionary definition for real estate is as follows: *Property in the form of land or buildings.*

Commercial real estate is utilised for almost all aspects of everyday life, including living through all stages of our lifecycles, desk working, shopping, moving physical goods, storing physical goods and items, and providing space for industrial activity. Commercial real estate can be directly held by owner-operators, who use the space for their own business needs and are thus fully responsible for all aspects of its management and upkeep or leased by investors to occupiers in return for rental income. Direct real estate capital markets provide the forum in which direct assets and portfolios are bought and sold by investors.

DOI: 10.1201/9781003298564-4

3.1.2   Ownership types

When they own property on a freehold basis, investors have outright ownership and full title over the property and land in question. When real estate is owned on a leasehold basis, the investor pays rent to the freehold owner for a defined period, typically a long-term (50–100 years+) period. An example of where leasehold ownership is more commonplace is directly on or adjacent to infrastructure facilities such as airports and ports. In these instances, the authorities seek to retain enhanced control over the land and associated properties given their critical use. In certain markets, most commonly in dense urban metropolitan areas such as Hong Kong, real estate can be owned on a strata title basis. A strata ownership investor owns a specific floor or pro rata area of the reference property, with the investor(s) having obligations and rights over the common areas (e.g., lobbies), structural maintenance and other aspects, such as facade signage.

3.1.3   Lease structures

Commercial real estate investors rent their properties to occupiers for fixed periods via a lease contract. They economically benefit from this income stream as well as potential capital appreciation over time. Lease contracts dictate the division of management responsibilities and expenses, with the owner typically accountable for all structural maintenance. Lease structures can include a proportion of or be fully based on profit or revenue sharing. For example, turnover-based leases are relatively common for certain retail property types. Investors can also engage specialist operators to manage assets on their behalf under management contracts. By way of example, these are prevalent in the hotel sector. In instances where investors lease on this basis, the ability for rent to be covered by the profits generated from activities within the asset is a key investor consideration using metrics such as EBITDA coverage (EDITDA-to-rent ratio).

3.2   Commercial real estate sectors

3.2.1   The "traditional" sectors

Historically, direct real estate capital markets have been dominated by what are termed the" traditional" real estate sectors, namely the office, retail, industrial and multifamily. Figure 3.1 provides an overview of the main property types within these sectors, their key features, and the occupier demand drivers. These are the most prevalent commercial real estate types used by occupiers and represent the bulk of institutional investor portfolios. Figure 3.3 highlights that for US institutional investors, as shown by the NCREIF National Property Index ("NCREIF-NPI") at the end of 2021, these sectors represented approximately 96% of direct commercial real estate portfolios.

Office, retail, and industrial assets generally entail a relatively low number of occupiers to manage which is in contract to the multifamily sector, making these easier to manage on a day-to-day basis.

3.2.2   The broadening sector menu

In recent years, several "alternative" or "other" sectors have seen their capital markets develop significantly and this has naturally been in lockstep with investor interest. Certain capital markets have already seen greater representation of these property types. At the end of 2021, the market capitalisation of US equity REITs was 41% allocated outside of

	Description	Lease Length / Stay Time	Operating Metrics	Demand Drivers	Structural trends
Office	• <u>CBD</u>: located in the main central business areas in major cities. • <u>Fringe-CBD</u>: in-fill and often central locations that are adjacent to the main CBDs. • <u>Suburban</u>: located in non-urban residential orientated areas • <u>Business parks</u>: standalone office parks/complexes typically located away from dense urban or suburban areas	3-10+ years	Rental rates Vacancy	Economic growth Office-based business and financial services employment for major CBDs	(-) Impact of work-from-home (-) Increasing densification from flexible use
Industrial	• <u>Logistics/distribution</u>: properties or "sheds" that are used for the movement of physical goods. Accessible for heavy goods vehicles and located in proximity to key transport networks. • <u>"Last mile" distribution</u>: smaller distribution facilities located in dense urban areas enabling the final stage of delivery of physical goods. • <u>Multi-let industrial</u>: estates with multiple units typically let to multiple occupiers for industrial and light manufacturing use. • <u>"Flex"</u>: typically, smaller single-storey industrial buildings accommodating a higher (25%+) office usage depending upon tenant requirements	3-10+ years	Rental rates Vacancy	Economic growth Technology	(+) E-commerce
Retail	• <u>Shopping centres/malls</u>: large and typically indoor buildings with multiple retail units available to occupiers. There is often food and beverage provision. • <u>Retail parks/strip centres</u>: outdoor retail centres with multiple units organised in rows. These can provide a mix or be focusedon grocery, fashion and/or bulky goods (e.g.,DIY) retailers. • <u>Supermarket/grocery anchored</u>: standalone grocery stores that could include an adjoining parade of unit shops. • Unit shop: standalone retail units typically located on high streets	3-10+ years 10-20+ years for supermarkets	Footfall Turnover Health ratios	Economic growth Consumer expenditure	(-) E-commerce
Multifamily	• <u>Multi-family</u>: purpose-built rental housing accommodation providing apartment units for multiple households. These can have high amenity provision (e.g.,concierge services and gyms) or provide more affordable accommodation	12-24+ months	Rental rates Vacancy	Personal income Household formation	(-/+) Demographics
Hotel	• <u>Full service</u>: hotel providing a range of services and facilities, such as in-house food and beverage, and can be luxury, upscale or mid-market in terms of price and quality. • <u>Limited service</u>: typically, "room only" mid-market or budget offering	1-14 nights when operated. 10+ years management contracts	Average daily rate ("ADR") Revenue Per Room ("RevPAR") = ADR * Occupancy	Economic growth Disposable income	(-) Covid Pandemic

Figure 3.1 Traditional commercial real estate sectors overview.

Sector	Description	Lease Length / Occupier Stay Time	Operating Metrics	Demand Drivers	Structural trends
Storage	• Self-storage: industrial properties configured for customers to rent storage units. • Cold storage: industrial and logistics properties facilitating the storage and movement of goods that need to be preserved in cold temperatures	6 months -10+ years	Rental rates Vacancy	Economic growth Technology	(+) E-commerce
Data Centres	• Colocation facilities: typically, smaller facilities where either space or usable capacity is leased to multiple customers. A range of services such as physical hardware maintenance can also be provided. Capacity requirements vary significantly but will be no more than one megawatt ("MW") • Hyperscale facilities: larger sites servicing capacity requirements from one MW to 50 MW+. Hyperscale customers include Amazon DWS and Microsoft	1-5+ years colocation 5-10+ years for hyperscale customers	Energy efficiency Rent per kW/MW	Economic growth Consumer expenditure	(+) Technological growth
Student Housing	• Student housing: purpose-built rental housing accommodation students located on or in proximity to universities and other higher education facilities. These can be provided in dormitory or self-contained unit formats	9-12 months	Rental rates Term time occupancy	Student funding University rating	(+) Demographics (+) Increasing higher education participation
Senior Housing	• Retirement housing/independent living: purpose-built facilities providing services such as meal provision, cleaning and providing social events/forums for elderly residents. • Care Homes: high acuity care provision including assistance with all aspects of everyday life for elderly residents. Can include specialist care provision for conditions such as dementia	24-36 months	Rental rates Vacancy	Ageing population growth Personal incomes & government funding	(+) Demographics – ageing populations
Healthcare	• Medical office: properties accommodation medial activities such as doctors' surgeries, clinics,and dentists • Hospitals: properties providing the facilities for surgery and the treatment of injuries and illness including their recovery	5-10 years+ for medical office Leased hospitals have 10-20+ year terms	Rental rates Vacancy	Personal incomes & government funding	(+) National healthcare expenditure

Figure 3.2 "Alternative" commercial real estate sectors overview.

the traditional categories in contrast to the private US market, which had an equivalent 4% exposure. REITs are well suited to these property types given they can be staffed with specialist internalised management capabilities whereas generalist real estate investors often look to outsource more operationally intensive activities through management contracts or other agreements.

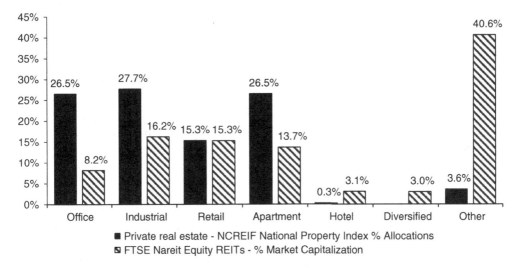

Figure 3.3 US private and public equity real estate allocations at end of 2021.

Sources: NCREF, ww.reit.com; authors' calculations.

As shown previously, these property types have unique attributes, such as their specialised use and demographically or other structural trend-driven demand factors. A sample of these are highlighted in Figure 3.2. A number of these can be considered as being "adjacent" to the traditional sectors, such as self and cold storage to more conventional industrial properties. However, they are typically more operational intensive than traditional sectors, often requiring specialist management expertise due to a greater degree of service provision. This is well illustrated by senior housing, where high acuity care provision is required.

3.2.3 *Asset quality profiles*

All commercial real estate can vary in terms of quality, which can be reflected by location, specification, physical condition and or income profile. Direct capital markets reference prime or Grade A definitions that typically represent modern, best-in-class properties in the best locations leased to creditworthy tenants. Secondary or tertiary locations are those outside of the main urban or less desirable areas in terms of physical geography, socio-economic or demographic attributes. In addition to location, secondary or tertiary assets can also be out-of-date, poorly configured, in need of structural repairs/upgrades or aesthetically unappealing. Increasingly, the environmental and sustainability characteristics of assets are having an important bearing on their quality. Organisations such as BREEAM[1] provide asset sustainability ratings that guide investors and occupiers.

3.3 Size of the direct real estate capital markets

3.3.1 *Market universe estimates*

Unlike publicly owned securities, where aggregate market capitalisation is a "live" observable metric, the size of direct real estate capital markets has to be estimated. This is because not all commercial real estate is held by investors and even assets that are held for investment purposes are not necessarily valued or do not have their performance measured periodically. The commercial real estate market universe can be defined in three ways: how much is held by investors, what could be held by investors, and what is transacted by investors.

MSCI, as reported in Hariharan et al. (2021), estimates the global market size of stock held by investors in 32 countries at $10.5 trillion as of the end of 2020. MSCI defines this estimate as representing "the aggregation of real estate assets that meet all of the following conditions: They are held as investments for the purposes of delivering a mix of income and capital returns; They are professionally managed for the achievement of these purposes, either by the beneficial owners or by third-party management businesses; They are structured as investment interests within portfolios". This measure excludes smaller landlords with less than 100 million of invested assets, owner-occupied portfolios, and municipal/social housing.

PGIM (2021) estimated that the total size of the global commercial real estate market that could be held by investors was at $31.1 trillion as of the end of 2020. PGIM quantifies this measure for 55 countries, which are deemed to have "functioning" commercial real estate investment markets, and it is based on country GDP and GDP per capita data. They cite several factors driving the difference between their estimate of what is investable to MSCI's invested universe, including regulatory environments, legal frameworks, and cultural practices. Ultimately, these factors relate to the openness and sophistication of broader capital and financial markets, with countries such as Hong Kong, Singapore, Sweden, and the UK having large invested markets relative to the size of their GDP.

As shown in Figure 3.4, the largest 20 countries represent approximately 87% to 95% of the estimated global market universe, and the top ten, 74 to 82%. In both instances, the US represents by far the single largest commercial real estate market, at approximately one-third. China is the major market where there is the most discrepancy in terms of share, with the PGIM investable estimate at 12% versus 6% for the MSCI invested market size. This can be attributed to the relatively limited openness of real estate capital markets in China when contrasted with Western economies.

There has been increasing transparency relating to recorded direct real estate transaction volumes. This can be considered something of a more practical market universe size estimate, as it reflects actual activity that investors could, at least theoretically, mirror. Real Capital Analytics (RCA) has logged the number and value of traded assets on a global basis. Over the past five years to the end of 2020, global volumes have averaged $1.6 trillion p.a. and were $1.5 trillion in 2020. Whilst investment activity in 2020 was disrupted by the impact of the COVID-19 pandemic, it represented over 14% of the MSCI investment universe estimate.

3.3.2   *Why understanding market universe size is important*

Understanding the size of direct real estate capital markets is important for commercial real estate investors embarking on regional or global programmes. These estimates help frame strategic real estate portfolio allocations, on which tactical allocations and/or investor-specific considerations, can be overlaid. At a geographic or sector level, they can also inform direct investors on market depth and liquidity profile. Finally, they also inform commercial real estate's position within broader multi-asset portfolio allocation frameworks.

Largest 20 Countries	MSCI Invested Market Size	Allocation %	Cumulative Allocation	Largest 20 Countries	PGIM Investable Market Size	Allocation %	Cumulative Allocation
United States	3,651	34.8%	34.8%	United States	9,370	30.1%	30.1%
Japan	940	9.0%	43.8%	China	3,821	12.3%	42.4%
UK	769	7.3%	51.1%	Japan	2,277	7.3%	49.8%
Germany	684	6.5%	57.7%	Germany	1,741	5.6%	55.4%
China	668	6.4%	64.0%	UK	1,560	5.0%	60.4%
France	500	4.8%	68.8%	France	1,210	3.9%	64.3%
Canada	364	3.5%	72.3%	Italy	870	2.8%	67.1%
Hong Kong	356	3.4%	75.7%	Canada	761	2.4%	69.5%
Australia	348	3.3%	79.0%	South	747	2.4%	71.9%
Switzerland	324	3.1%	82.1%	Australia	623	2.0%	73.9%
Sweden	294	2.8%	84.9%	Spain	565	1.8%	75.7%
Netherlands	210	2.0%	86.9%	Brazil	506	1.6%	77.4%
Singapore	194	1.9%	88.7%	India	462	1.5%	78.9%
Italy	146	1.4%	90.1%	Netherlands	407	1.3%	80.2%
Spain	128	1.2%	91.4%	Russia	407	1.3%	81.5%
South Korea	105	1.0%	92.4%	Mexico	388	1.2%	82.7%
Finland	101	1.0%	93.3%	Hong Kong	356	1.1%	83.9%
Denmark	84	0.8%	94.1%	Switzerland	332	1.1%	84.9%
Norway	67	0.6%	94.8%	Singapore	322	1.0%	86.0%
Belgium	66	0.6%	95.4%	Sweden	294	0.9%	86.9%

Figure 3.4 Invested vs. investable market size estimates at the end of 2020.

Sources: MSCI, PGIM, authors' calculations.

3.4   Participating in direct real estate capital markets

3.4.1   Market-makers and service providers

There are various commercial real estate investor profiles with wide-ranging motivations and requirements, but assisting these investors in all aspects of their participation in direct real estate capital markets are third parties who facilitate transactions and advise on transactions. All these groups are remunerated by fees, and these may be partly to fully payable depending on whether an investor successfully completes a transaction. An investor's level of experience, market knowledge and resourcing will drive their reliance on these various groups, which are as follows:

- *Real estate brokers or investment agents:* These are intermediaries of commercial real estate transactions, who introduce assets to buyers when mandated by vendors. They also manage the transaction processes described in Section 3.3.2.
- *Other intermediaries:* These cover a range of groups but most notably banks and other financial advisers when real estate is being transacted through corporate entities and other holding structures.
- *Property advisers:* These include the large full-service providers such as CBRE and Jones Lang LaSalle (which also have significant brokerage practices) through to more

niche or focused service providers who advise on all aspects of market, financial, physical, and ESG-related due diligence activities.

- *Lawyers:* They advise on all legal aspects, including title/ownership considerations and documentation such as leases and transaction contracts. Legal advisers also ensure that transactions are executed in a legally binding manner and that due process is followed.
- *Accounting firms and tax advisers:* These advisers provide tax advice pertaining to an investor's income or capital gains tax liability. Where direct real estate is transacted through ownership structures, such as corporate entities, they also provide financial due diligence services and ensure correct financial proceeds and fund flows are processed.
- *Operating partners/third-party asset managers:* They assist investors wishing to fully outsource or have ongoing asset management activities and also introduce transactions.

3.4.2   *How do investors participate in direct real estate capital markets?*

Directly held real estate is accessed by investors through private market transactions. They can originate opportunities through the following processes:

- *Marketed:* Sales processes are managed by real estate brokers and other intermediaries, where assets or portfolios are introduced to a broad group of potential buyers. These more transparent processes are typically well-defined, with multiple bidding rounds based on the standardised information provided. Vendors seeking to maximise their proceeds will often use this process. Intermediaries coordinate these processes, guiding bidders and advise vendors on best execution. Best execution does not necessarily mean the highest bid wins; instead, other considerations such as buyer reputations, funding status and profile can have a material bearing.
- *Auctions:* These are marketed processes where buyers bid at a set time in an open forum. A sealed bid process, where "best and final" bids are provided in writing at a set time, can be used, with bidders unaware of the pricing provided by one another. These processes are managed and adjudicated by auctioneers.
- *Limited marketing:* These less transparent processes are typically intermediated with introductions to a relatively small number of prospective buyers. The rationale for vendors using this approach could be that specialist buyers who need specific operational expertise to assess the reference asset(s) are required or they have time constraints and so focus on selling to well-capitalised and sophisticated buyers who can close expediently.
- *Off-markets:* This is where no marketing process is followed. Instead, sellers bilaterally negotiate with a single or small number of prospective buyers with or without an intermediary. Vendor preferences drive whether this route is followed, and their rationales can in practice be wide-ranging, including views that they have already received the best pricing advice and that there is a need to be discreet for reputational purposes.

3.5   Creating new investible stock

3.5.1   *Accessing stock through development*

The above processes can be followed for standing individual properties or portfolios, but they can also be used to access newly constructed properties from developers. Investors can either:

1 Participate alongside the developer in all or the early phases of a development project from the land purchase, to obtaining the requisite planning permits, and funding construction, and can then either choose to own the completed asset or sell.
2 Acquire the permitted land and fund the construction phase – a "forward funding" transaction.
3 Or agree to acquire the completed asset before the commencement or during the construction phase via a "forward purchase" transaction.

The first route entails the most speculative and higher-risk activity due to the additional inherent uncertainty associated with the planning and permitting processes. This uncertainty includes timing considerations, working with multiple parties, and involving planning authorities on viability and design. If the requisite permits process cannot be obtained, this can render the land value worthless, or alternative uses will need to be explored.

In the case of 1) and 2), developers are often paid fees to manage and oversee the project. In the case of 1), they can participate in a profit-sharing arrangement. Investors and developers can mitigate construction cost risk by fixing build prices with third-party contractors, who, in turn, bear any unexpected price increases relating to labour or materials. The contractors are then obligated to build and fit out the asset to the appropriate specification within a set timeframe. Penalties will typically apply in the event of delays or other sources of non-compliance with their contractual obligations. This risk can also be insured; however, understanding the capabilities and balance sheet strength (credit profile) of third-party contractors are key considerations.

3.5.2 *Accessing stock through sale and leaseback transactions*

Following the origination processes outlined in Section 3.3.2, investors can also access direct assets through sale and lease back transactions with corporate owners. Corporations often provide investors with long-term leases of 10–20+ years and fully operate the assets. Their rationale for selling their ownership interests in what typically ranges from being typically important to critical for their businesses is the ability to secure the long-term use through the lease whilst releasing liquidity, which can then be deployed to more profitable uses. Investors benefit from long-duration income streams, with limited to no direct management. Key considerations include the creditworthiness of the corporate counterparty and their ability to service the rental payments, the importance of the asset to the operations, and alternative use value in the event of default or lease expiry. Well-developed direct real estate capital markets exist for such transactions and assets with this profile, for example, the net lease market in the US.

Quiz questions

1 Provide examples of where leasehold ownership is more commonplace.
2 What is strata ownership?
3 What are the traditional sectors of institutional real estate ownership?
4 Why is it important to understand the estimates of the size of different real estate markets?
5 How is new investible stock created?

Discussion questions

- How has the role of traditional sectors in an institutional portfolio evolved post-2017?
- How would an institution gain access to suitable investible stock?
- How do you expect the relative investible size of sectors and countries to change over the next ten years?

Research/Dissertation topics

- Are current benchmarks representative of the investible universe for institutional investors?
- What are the risks for institutional investors of switching from traditional sectors to those which are more operationally intensive?

Note

1 www.breeam.com

References

A Bird's Eye View of Real Estate Markets, PGIM Real Estate, October 2021.
Real Estate Market Size 2020, Annual Update on the Scale of the Professionally Managed Global Real Estate Investment Market, Hariharan, G.G., Patkar, R., and Neshat, R., MSCI, August 2021.

4 Analysing a direct real estate investment

Objectives

At the end of this chapter, you will understand how to:

- Project operating cashflows for direct real estate investments.
- Consider relevant capital expenditures.
- Estimate exit values using various market yields or cap rates.
- Project total returns and see how they are impacted by key variables.

4.1 Assessing and projecting income

4.1.1 Lease contracts

As summarised in Chapter 3, commercial real estate investors lease their properties to occupiers and receive contractual rental income stream in return. Leases are legal agreements with a range of provisions, although there are high degrees of uniformity within specific geographies and property types. Figure 3.1 highlights the typical lease lengths for a range of property types. For most "traditional" property types, real estate investors directly lease their assets on a long-term basis. But this is not the case for more operationally intensive segments where occupier stay-times are shorter, for example, multifamily healthcare, hospitality, or self-storage, where anything from one-day to one-year lease lengths prevails. Here investors may elect to operate these property types directly to occupiers or lease to creditworthy operators who, in turn, may lease to end-occupiers. These leases are then materially longer than those for the underlying end-occupiers and this reduces the execution risk for investors, with commensurately adjusted return expectations.

Commercial real estate leases normally have the following legal terms and stipulations:

- *Term:* start and end dates.
- *Premises:* the specific area a tenant occupies.
- *Rent Reviews:* interim rental income "resets" to market levels on stated dates; these are typically upwards only.
- *Contractual Uplifts:* any fixed percentage or referenced (e.g., inflation-linked) uplifts, which are typically annual from the start of the lease.
- *Tenant and Owner Expense Obligations:* the operational cost items (e.g., property managers, maintenance expenditures) are paid by the occupier or property owner.
- *Restrictions:* these apply to the use of space, for example, the ability to sub-lease and make physical alterations, and

DOI: 10.1201/9781003298564-5

• *Tenant Credit:* any occupier deposits or guarantees made (e.g., from the parent company owners). In the event of occupier default on their rental payment obligations, there are remedies available to the owner and an ultimate course of action to take back the premises.

Start Date   31-Mar-22

Unit	Lease Start	Lease End	Space Leased - Metres	Remaining Term - Years	Rent Per Square Metre	Market Rental Value ("ERV")	Rent Payable	Market Rent	Rent Escalations	Escalation Basis
1	31-Mar-19	31-Mar-29	10,000	7.0	20.0	25.0	800,000	1,000,000	Yes- Annual	CPI
2	31-Mar-19	31-Mar-24	10,000	2.0	22.5	25.0	900,000	1,000,000	No	N/A
Total			20,000	4.5	21.3	25.0	1,700,000	2,000,000		

Figure 4.1 Illustrative tenancy schedule.

Source: Authors.

Commercial real estate leases may also provide for what are generally tenant-favourable extensions or break options. The relevant dates, deadlines and terms would be contractually stipulated. Tenants may also have the rights to expand the space that they occupy. This may include take-up rights in the form of a right-of-first-refusal ("ROFR") where other occupiers either express interest or the space becomes available.

4.1.2 *Illustrative tenancy schedule*

An illustrative hypothetical tenancy schedule is shown in Figure 4.1. This provides an example of the information that investors typically receive when assessing a potential new direct real estate acquisition.

For each individual tenant, a lease term, the rent payable, and an assessment (typically from an appointed appraiser) of the prevailing market rental level or "ERV" apply if a new lease is struck on market terms. Using this information and other key lease terms provided for their due diligence, investors can make their own NOI and cashflow projections, but will need to make several assumptions to do so.

4.1.3 *Projecting NOI*

Real estate investors receive rental income from occupiers of their properties and will also incur operating expenses for managing these assets, with the net of these being NOI. These operating expenses include ongoing costs and fees such as insurance, local property taxes, property management (including rent collection), maintenance expenditures and utilities. Depending upon the market convention according to the property and/or geography, these j will be payable to a greater or lesser extent by real estate owners. They can vary considerably. At the extreme, there are "triple net" leases or NNNs in US market parlance, where the tenant is responsible for all operating expenses as they fully operate the property, as per the leased operating real estate example provided above in Figure 4.1. These have long-term, 10–20+ year leases often with annual inflation-linked or fixed percentage escalations.

NOI is calculated by the following steps:

Potential rental income: fully leased property at market rates or ERV
+/– Leased rent versus market

- *Vacancy*
- *Credit losses (tenant defaults)*
+ *Other income*
= **Gross income**
- *Operating expenses*
= **Net operating income**

Figure 4.2 shows an illustrative six-year NOI projection or "pro-forma NOI statement" for the hypothetical tenancy schedule in Figure 4.1.

For the purposes of this worked example, the items and illustrative analysis in this chapter are calculated on an annual basis for simplicity's sake, but investors should consider projections on monthly or quarterly periods as they deem appropriate. Firstly, to project gross income, market rental growth and inflation forecasts are required. Investors could form their own expectations of these variables using trend analysis or econometric modelling or use third-party provided or consensus expectations where these are available. The lease for Unit 1 expires beyond the six-year horizon of the pro-forma and so, absent any tenant default assumptions, the rental income escalates annually by the three percent per annum projection assumption made for inflation (CPI).

For Unit 2, the existing lease expires at the end of the second year of the projection period, and it is assumed that this occupier will vacate these premises. In this scenario, we then assume that the space will take 12 months to lease and then, as part of an incentive package for a prospective new tenant, the first 12 months of the new lease will be free for this tenant. This is termed a "rent-free period". Investors can also opt to model this aspect

	Item / Variable	Row	Current Mar-22	Year 1 Mar-23	Year 2 Mar-24	Year 3 Mar-25	Year 4 Mar-26	Year 5 Mar-27	Year 6 Mar-28
	Market Rental Growth	1	2.0%	2.0%	2.0%	2.0%	2.0%	2.0%	2.0%
	Inflation - CPI	2	3.0%	3.0%	3.0%	3.0%	3.0%	3.0%	3.0%
	Rental Income								
	Tenant 1								
	Market Rent / ERV	3	1,000,000	1,020,000	1,040,400	1,061,208	1,082,432	1,104,081	1,126,162
	Rent Payable	4	800,000	824,000	848,720	874,182	900,407	927,419	955,242
	% vs Market	5	*80%*	*81%*	*82%*	*82%*	*83%*	*84%*	*85%*
	Tenant 2								
	Market Rent / ERV	6	1,000,000	1,020,000	1,040,400	1,061,208	1,082,432	1,104,081	1,126,162
	Rent Payable	7	900,000	900,000	900,000	0	0	1,061,208	1,061,208
	% vs Market	8	*90%*	*88%*	*87%*	*0%*	*0%*	*96%*	*94%*
	Total Gross Rental Income	9	**1,700,000**	**1,724,000**	**1,748,720**	**874,182**	**900,407**	**1,988,627**	**2,016,450**
	Operating Expenses								
6.0%	Local Property Taxes	10	102,000	103,440	104,923	52,451	54,024	119,318	120,987
5.0%	Maintenance / Repairs	11	85,000	86,200	87,436	43,709	45,020	99,431	100,822
4.0%	Property Management Fees	12	68,000	68,960	69,949	34,967	36,016	79,545	80,658
50,000	Insurance	13	50,000	51,500	53,045	54,636	56,275	57,964	59,703
25,000	Utilities	14	25,000	25,750	26,523	27,318	28,138	28,982	29,851
10,000	Professional Fees	15	10,000	10,300	10,609	10,927	11,255	11,593	11,941
7,500	Other Expenses	16	7,500	7,725	7,957	8,195	8,441	8,695	8,955
	Total Operating Expenses	17	**347,500**	**353,875**	**360,441**	**232,204**	**239,171**	**405,527**	**412,917**
	% Gross Rental Income	18	*20%*	*21%*	*21%*	*27%*	*27%*	*20%*	*20%*
	Net Operating Income	19	**1,352,500**	**1,370,125**	**1,388,279**	**641,977**	**661,236**	**1,583,100**	**1,603,533**

Figure 4.2 Illustrative NOI projection.

on a probability-weighted basis. In this example, we are assuming a 0% renewal probability weight. But should an investor have varying degrees of confidence on whether a tenant will renew their lease, they can assign an appropriate weighting. This would then be used to adjust vacancy or "down-time" periods or income projections. There is a 24-month period when no rental income is received by the investor from this space. The new lease is also not assumed to benefit from annual inflation-linked escalations. So, the projected market rent at the end of year three, the renewal date, is the pro-forma gross income forecast.

Operating expenses can be calculated in two ways. Firstly, they can be estimated on a fixed percentage of gross income, and this may be the calculation basis for items such as property taxes. Other operating expenses will be billed on a monetary basis, and we consider it prudent for investors to grow these fixed costs over time. For example, in Figure 4.2, the fixed-cost operating expenses in rows 13 to 16 are grown by projected inflation of 3% per annum. Due to the presence of both variable and fixed operating expenses, during the period when Unit 2 is not providing rental income, the operating expenses to gross rental income ratio, the "NOI margin", steps up meaningfully.

4.2   Capital expenditure and pricing

4.2.1   *Capital expenditure and obsolescence*

In addition to operating expenses, investors also have to make provision for capital expenditures for repairs, upgrades, or renovations and/or for potential expansions. This is to ensure that longer-term property performance is not unduly impacted by physical obsolescence or other sources of depreciation. To illustrate the difference between operating and capital expenditures we can use the plant and machinery in a building. This will require ongoing servicing, which could include the replacement of components. This is a maintenance expense. But when it requires a substantial refit or complete replacement, that becomes a capital expenditure item.

Depreciation in a specific building is driven by several factors and varies across property types and geography. The rate of depreciation will be inherently lower when the land value forms a greater share of the value of a property to asset value, for example, for prime high street shops or retail units where the location itself is key rather than the building itself. Other considerations include a building not having the requisite sustainability profile, a particularly pertinent issue in the current environment, or whether the specification can accommodate the latest technology or modern working practices. All buildings will ultimately suffer depreciation over time and investors should be prudent in provisioning for this in the form of offsetting expenditure, whether that be partial or in full, in their assessments of property investments.

To highlight the impact of depreciation on commercial real estate performance, a study by Bokhari and Geltner (2018), utilising over 100,000 transaction datapoints for the US, found an average depreciation rate of 1.5% p.a. of total property value, inclusive of land. This varied from approximately 1.8% p.a. for newer properties to 1.1% p.a. for those older than 50 years. There was significant variation across metropolitan markets, with the large east and west coast metros where land is more constrained having average lower rates, highlighting land value's considerable bearing on this phenomenon. Depreciation was largely shown to be reflected in reduced NOI versus pricing (cap rate) adjustments. Other academic and industry studies on this topic for further reading include Crosby et al. (2011), Fisher et al. (2005), and IPF (2011).

4.3 Entry and exit capitalisation

In estimating prospective returns for a direct real estate investment opportunity, investors need to determine its price at both acquisition and planned disposition. This is typically accomplished using a cap rate or yield to capitalise income. Market conventions differ on the precise definitions, but starting with financial theory, investors can adapt Gordon's (1962) growth model to estimate a commercial real estate yield as follows:

Commercial real estate yield $= RFR + RP - G + D$

Where:

- RFR is a reference risk-free rate, typically a government bond yield or other interest rate.
- RP is a risk premium reflecting the incremental total return that an investor should require given the risk entailed.
- G is expected growth.
- D is assumed deprecation.

Direct real estate risk premiums are discussed in greater detail in Chapter 5, but the formula states that the cap rate or yield that investors adopt should be the unleveraged required return (RFR + RP) being sought less rental income growth expectations net of depreciation.

In practice, a range of definitions are adopted by investors, with local market conventions varying considerably in terms of the yield estimated given a representative direct real estate investment. These differences depend on the definition of rental income (gross or net) that is adopted for the numerator and whether market-rate acquisition costs are incorporated in the denominator. Figure 4.3 provides several of the most widely utilised definitions for the acquisition of the illustrative property example we have been referencing throughout this chapter and identifies which items are included for each.

Firstly, we assume a US convention cap rate of 6% as the purchase price basis with the other definitions estimated using the 22.8 million acquisition price in their denominators. Initial yields, both gross and net, are more frequently referenced in Europe, with for example the German market focused on the gross initial yield or its inverse gross multiple, whereas the net initial yield is the typical reference point in the UK. It should be noted that net initial yield is sometimes calculated using gross income as the numerator and so investors should be aware of which calculation is being used when evaluating this metric.

Items	Amounts	Cap Rate	Trailing Cap Rate	Gross Initial Yield	Net Initial Yield	Gross Reversionary Yield	Net Reversionary Yield
		6.00%	5.92%	7.44%	7.16%	8.76%	8.42%
Included in Numerator							
Current Gross Income	1,700,000			✓			
Current ERV	2,000,000					✓	✓
Current NOI	1,352,500		✓		✓		
Year 1 Projected NOI	1,370,125	✓					
Included in Denominator							
Purchase Price	22,835,417	✓	✓	✓	✓	✓	✓
Transaction Costs - 4% of Price	913,417				✓		✓

Figure 4.3 Cap rate and yield definitions.

Source: The authors.

Reversionary yields guide investors on what a property could yield could "revert" to if it were fully leased at prevailing market rents. As can be seen from Figure 4.3, international investors need to interpret yields across markets carefully to ensure that comparisons are being made on a like-for-like basis, given the significant variation in these estimates. Using the same set of metrics on a projected basis, investors can easily project exit pricing for the direct real estate investments that they are considering.

4.4   Putting it all together – projecting total returns

4.4.1   *Illustrative direct real estate investment cashflows*

To analyse the prospective returns for a direct real estate investment, investors need to assemble all the variables and metrics discussed in this chapter, to create an aggregate cashflow. As illustrated in Figure 4.4, in addition to NOI, the cashflows need to include transaction costs, which we assume are 4% of purchase price and 2% of sales price, leasing fees (here assumed at 10% of newly leased gross rental income) and budgeted capital expenditures. Entry and exit pricing are determined using US cap rates based on 12-month forward NOI with an entry rate of 6% and an exit rate of 5.5%. When combined, it is possible to calculate prospective performance, with Figure 4.4 showing the calculated IRR and profit-on-cost or the equity multiple, which is calculated by dividing the sum of all positive cashflows by the sum of all negative cashflows. In this example, a five-year hold period is assumed for the investment, but investors can adjust this to suit their own preferences.

4.4.1.1   *Sensitivity analysis*

Having created a "central" or "base case" projected total return assumption, investors can then assess the sensitivity of this to changes in the underlying assumptions for the variables considered. This can help investors to understand the viability of opportunities under a range of scenarios, whether these be driven by market movements, for example, economic variables or asset-specific considerations, such as variance in the level of operating and

Item / Variable	Row	Current Mar-22	Year 1 Mar-23	Year 2 Mar-24	Year 3 Mar-25	Year 4 Mar-26	Year 5 Mar-27
Net Operating Income	19		1,370,125	1,388,279	641,977	661,236	1,583,100
Purchase Price	20	(22,835,417)					
Transaction Costs	21	(913,417)					
Leasing Fees	22						(106,121)
Capital Expenditure	23		(150,000)	(150,000)		(300,000)	(50,000)
Projected Exit Price	24						29,155,137
Sales Costs	25						(583,103)
Investor Cashflow	**26**	**(23,748,833)**	**1,220,125**	**1,238,279**	**641,977**	**361,236**	**29,999,014**
IRR	**27**	**7.6%**					
Profit-On-Cost	**28**	**1.41x**					

Figure 4.4 Projected unlevered cashflows and total returns.

Source: Authors' calculations.

IRR Sensitivity		Market Rental Growth				
		-2.0%	0.0%	2.0%	4.0%	6.0%
	6.5%	3.2%	3.8%	4.5%	5.1%	5.8%
	6.0%	4.7%	5.3%	6.0%	6.6%	7.3%
Exit Cap Rate	5.5%	6.3%	7.0%	7.6%	8.3%	9.0%
	5.0%	8.2%	8.8%	9.5%	10.2%	10.9%
	4.5%	10.2%	10.9%	11.6%	12.3%	13.1%

Profit-on-Cost Sensitivity		Market Rental Growth				
		-2.0%	0.0%	2.0%	4.0%	6.0%
Exit Cap Rate	6.5%	1.16x	1.19x	1.22x	1.26x	1.30x
	6.0%	1.23x	1.27x	1.31x	1.35x	1.39x
	5.5%	1.33x	1.37x	1.41x	1.45x	1.50x
	5.0%	1.44x	1.48x	1.53x	1.58x	1.62x
	4.5%	1.58x	1.63x	1.68x	1.73x	1.78x

Figure 4.5 Projected total returns sensitivity – exit yield and rental growth projections.

Source: Authors' calculations.

capital expenditures budgeted. As such, this analysis often forms a key component of risk management processes and guides investors on the extent to which they are exposed to potential downside risks and upside potential from a financial perspective.

Figure 4.5 provides an example of such analysis and highlights the variation in the expected IRR and profit-on-cost across a range of market rental growth and exit cap rate assumptions. The performance projections from Figure 4.4 are highlighted. It should be noted that in this example, the market rental growth sensitivity will only affect Unit 2 due to its impact upon the released rental level, whereas Unit 1 will not be impacted over the five-year horizon. Also, as the cap rate is the adopted exit pricing basis, it will not account for projected market rental levels whereas a reversionary yield assumption would. This example highlights the need for investors to consider a range of yields and cap rates in their assessments of direct real estate opportunities.

Quiz questions

1 List the key factors that would drive the rental growth outlook for a commercial real estate asset.
2 List the attributes that could determine either the new leasing prospects or renewal probabilities for office space.
3 What is the difference in the calculation of gross and net initial yields?
4 Using the illustrative direct real estate investment example cashflow in Figure 4.4, what would be the IRR dilution of a 75% increase in the budgeted capital expenditures?
5 Using the illustrative direct real estate investment example cashflow in Figure 4.4, fill in the following sensitivity analysis matrix for a range of CPI growth and exit cap rate assumptions that will have an impact.

Discussion questions

- Do you think landlords and valuers provision sufficiently for depreciation in commercial real estate?
- How does the increased focus on sustainability ratings for buildings impact the required capital expenditure from asset owners?
- How much can exit yields expand at times of market turbulence and can this be accurately forecast?
- How much do commercial real estate operating expenses and capital expenditures vary across sectors and geographies?

Research/Dissertation topics

- How accurately do theoretical market valuations reflect transaction-based evidence across sectors and cycles?
- How do investors make provision for depreciation? Use a survey-based analysis.

References

Bokhari, S. & Geltner, D. (2018). Characteristics of depreciation in commercial and multifamily property: An investment perspective. *Real Estate Economics*, 46(4): 745–782.

Crosby, N., Devaney, S. & Law, V. (2011). Benchmarking and valuation issues in measuring depreciation for European office markets. *Journal of European Real Estate Research*, 4(1): 7–28.

Fisher, J.D., Smith, B.C., Stern, J.J. & Webb R.B. (2005). Analysis of economic depreciation for multifamily property. *Journal of Real Estate Research*, 27(4): 355–369.

Gordon, M.J. (1962). *The investment, financing and valuation of the corporation*. Irwin, New York.

IPF. (2011). *Depreciation of commercial investment property in the UK*. Investment Property Forum/IPF Educational Trust, London.

5 Valuing direct real estate assets

Objectives

At the end of this chapter, you will understand:

- The basics of valuations or appraisals.
- Market-based valuation estimates.
- How to estimate discount rates of required rates of return.
- How to use a discounted cashflow model to calculate investment worth.

5.1 Basics of valuation

The valuations of direct commercial real estate assets and portfolios, are determined by industry participants using two broad approaches:

1 *Market:* estimated by reflecting the cost or price at which an asset would trade for in the open market on an arms-length basis and
2 *Investment worth:* calculated by valuing the stream of future benefits, namely the projected income and capital returns. This is typically undertaken using a DCF model, with performance expectations and return requirements driving the calculations of value.

Because commercial real estate assets are heterogeneous and are traded infrequently, there is a range of independent valuation providers available who operate globally under industry and regulatory supervision. Crosby et al. (2018) detail the global industry bodies which provide best practices and standards to valuation professionals in their respective jurisdictions.

5.1.1 What are they used for?

Appraisals are central to the functioning of global real estate capital markets through their role in the financial system, the performance measurement of investments and investor assessments of existing and prospective investment opportunities. Valuations undertaken across property and owner types using the relevant approach are utilised for the following:

- Building insurance and reinstatement assessments.
- Reported balance sheet assets for publicly traded and privately held entities.
- Tax assessments by public agencies.

DOI: 10.1201/9781003298564-6

- Underwriting of commercial real estate mortgages and other debt finance to real estate entities.
- Basis for compulsory purchases made by public agencies.

Market valuations underpin performance measurement for both publicly traded and privately held entities such as REITs and private real estate funds. They are also used for regulatory purposes, including capital adequacy charges for insurance company investments. Assessments of worth using appraisals are used for investment decision-making purposes, whether for the periodic assessments of existing assets or when evaluating prospective purchases.

5.2   Market valuations of commercial real estate assets

5.2.1   Cost approach

This is the simplest approach for appraising a property. It states that the market value of a reference asset equates to the build cost of replacing it, inclusive of land. The two main ways that the cost approach can be applied are to estimate the cost to reproduce or replace an asset. To reproduce or reinstate an asset, one must simply estimate the replacement cost for a replica asset, providing the same occupier use and utility on the land and then deducting the depreciation incurred. For the replacement method, the cost of rebuilding the same size structure but on a modern basis using new materials, construction methods and designs.

Whilst the logic of a cost approach makes sense intuitively, the method does not directly account for the time and effort required to construct from the ground up or the cashflow potential of an asset, although investors can estimate these. Nonetheless, for investors, replacement cost estimates can be helpful yardsticks for value assessments, and in situations where one can purchase an asset below its estimated cost to either reproduce or replace, it signals potentially attractive value being available and vice versa.

5.2.2   Market comparable approach

Referencing relevant market-comparable price metrics is a straightforward way in which a reference asset can be valued. These metrics are also key inputs in a valuation model, particularly when independent third-party appraisals are conducted, which is discussed below. According to the 2020 IVS Global Standards 2020 and the RICS *Red Book* UK Supplement 2019, market value is defined as follows:

> *Market Value is the estimated amount for which an asset or liability should be exchanged on the valuation date between a willing buyer and a willing seller in an arm's length transaction, after proper marketing and where the parties had each acted knowledgeably, prudently and without compulsion.*

Market participants reference the information for recently sold properties that can be directly compared to a property being assessed. Considerations beyond property type, location, structural specification and, ever increasingly, sustainability profile include physical condition, building aesthetics, amenity provision, tenant profiles and creditworthiness. Widely referenced direct capital market sales metrics include customary yields or cap rates for a particular geography or sector (please refer to Figure 4.3 for the most commonly adopted definitions) and capital values on a local unit area basis, for example, per square metre or feet.

Depending upon the property type in question, there are also likely to be other valuation reference points. For example, hotel values or prices are often referenced on a per-key basis, or for residential blocks, unit prices are most widely cited. This approach can be thought of as a benchmarking exercise. The more directly comparable properties are, the more accurate any market-comparable valuation will be. This approach is dependent upon there being at least a reasonably functioning market, with a degree of observable pricing data points. However, like the cost approach described above, the market comparable approach does not account for the future stream of economic benefits that can be derived.

5.2.3 *Simple capitalisation approach*

Following the market comparable approach, relevant local market real estate yields or cap rates are often relied upon to value assets on a "spot income basis", using either existing and/or potential income. This may be the prevailing rental level or NOI of a reference asset, or a 12-month forward NOI expectation in the case of US cap rates. This method is most useful when appraising stabilised income-producing properties. It is commonplace to base this value on the estimated or established market rental levels of a property, and, in the case of over-rented properties, to effectively remove the value accruing to the over-rented portion of the contracted rental income.

Again, citing the 2020 IVS Global Standards 2020 and the RICS *Red Book* UK Supplement 2019, market rent is defined and used as follows:

> *Market Rent is the estimated amount for which an interest in real property should be leased on the valuation date between a willing lessor and a willing lessee on appropriate lease terms in an arm's length transaction, after proper marketing and where the parties had each acted knowledgeably, prudently and without compulsion.*

> *Market Rent may be used as a basis of value when valuing a lease or an interest created by a lease. In such cases, it is necessary to consider the contract rent and, where it is different, the market rent.*

5.3 Investment worth – DCF model-based approach

5.3.1 *Introduction*

In this section, we extend the worked example direct property cashflow from Chapter 4 to estimate investment worth by undertaking a DCF model-based appraisal exercise by calculating the net current value. Valuing by using cashflow projections of a property yields the expected income and gains (or losses) of a property. It can reflect specific considerations and potential asset management opportunities. Sensitivity analysis can also be undertaken in this approach to understand the key risk-return drivers underlying a valuation.

5.3.2 *Discount rates or required rates of return*

To undertake DCF valuations, investors or values need to assume a discount rate or required return that appropriately reflects the risk profile and other characteristics of the reference asset. This can then be used to calculate the current value of the projected cashflows. As per Gordon's growth formula summarised in Section 4.2.2, a required return can be built up using a reference risk-free rate (typically a government bond or similar fixed income security yield) and a risk premium representing an appropriate total return compensation spread for the risk entailed with the investment being considered.

A risk-free is commonly considered to be represented by a long-dated government bond yield for the country in which the asset is located, for example, a US Treasury yield for US assets. This can be estimated using current yield levels or projections from, for example, a consensus survey. Investors may also match the intended hold period for an asset with the relevant yield maturity.

To estimate the risk premium, one can use ex-post estimates by calculating the historical average performance spreads between total returns series and risk-free rates. Figure 5.1 provides such estimates for UK and US commercial real estate markets using annual performance and national 10-year government bond yields over several term timeframes through to the end of 2021. As can be seen, there is a degree of variation over long-term horizons but generally an approximate 3% p.a. is evident.

Whilst these estimates in Figure 5.1 are at a national all-property level, adjustments can and should be made for sectors assuming data is available. Figure 5.2 provides CAPM-based beta estimates for the major sectors based on a relevant direct real estate total returns series. As can be seen, offices represent a relatively higher risk whilst multifamily is relatively low. To put this in context, the 30-year UK 3.1% risk premium estimate would translate to equivalent risk premiums of approximately 3.4% (3.1% multiplied by 1.1x) for offices and 2.2% for multifamily, so you get a 1.2 percentage point lower required return for multifamily versus office.

Illustrative academic studies estimating commercial real estate market risk premiums to calculate discount rates and required returns include Breidenbach et al. (2006). This

	US	UK
30 years	3.4%	3.1%
35 years	2.7%	3.1%

Figure 5.1 Real estate total returns less local 10-year govt. bond yield, annual average p.a. to 4Q21.

Sources: MSCI, NCREF; authors' calculations.

Timeframe	US				UK			
	Retail	Office	Industrial	Multifamily	Retail	Office	Industrial	Multifamily
30 years	0.9	1.2	1.0	0.9	0.9	1.1	0.9	0.7
35 years	0.8	1.2	1.0	0.9	0.9	1.1	0.9	0.5

Figure 5.2 Real estate sectors' beta estimates to 4Q21.

Sources: MSCI, NCREIF, authors' calculations.

study provides more detailed risk premia estimates for sectors and metros in the US utilising estimated betas from equivalent REIT markets to inform their direct assumptions. Addae-Dapaah et al. (2014) employed a multifactor model approach to estimate risk premiums across a range of global markets and found considerable variation across markets and major property types.

This all relates to quantitative measures led by traditional asset pricing frameworks, but investors can also consider market size and liquidity, as well as general transparency levels. Although these can be challenging to quantify and translate into risk premium contributions, measures do exist. For example, Jones Lang LaSalle produces a global real estate transparency index that scores and ranks markets,[1] with the 2022 scores shown for the "Most Transparent" markets in Figure 5.3.

Whilst these estimates are at a market level, whether that be at a geographic or a sector level, or both, there is a need to account for property-specific factors. These include:

- Micro-location profiles, such as proximity to transport nodes.
- Building specifications or structure.
- Sustainability profile.
- Aesthetics.
- Amenity provision.
- Tenant credit profile(s).
- Duration of leased income.
- Significant pending lease events.
- Requirements for significant capital expenditure relating to upgrading or refurbishment works.

It is challenging to quantitively draw upon available data to estimate asset-specific risk premiums based on multifactor models, but investors can adopt other approaches to adjust market-level estimates. For example, a scorecard approach can be used to profile assets, with the resulting risk premium adjustments subject to minimum or maximum levels. The translation of these scores and their weightings into risk premium basis points is, however, based upon individual investor assumptions and preferences.

5.4   Putting it all together – an example DCF model valuation

In this section, the worked example direct property cashflow model from Chapter 4 is extended from projecting annually to quarterly over ten years to undertake a discounted cashflow model-based appraisal exercise. The appraisal or valuation is estimated by calculating the current value using a discount rate assumption. By valuing using the cashflow projections of a property, the expected income, and gains (or losses) from it can reflect specific considerations and potential asset management opportunities. This is typically undertaken using a DCF model-based approach, with performance expectations and return requirements driving estimates of value.

Figure 5.4 provides a snapshot summary output of the worked example in the accompanying Excel file for this chapter, where the full set of assumptions underlying these projections can be reviewed. A discount rate of 7% is assumed. It can be considered as being built up from a risk premium estimate of 3% p.a. from Figure 5.1 and a risk-free rate of 4%. This is used to calculate the current value of projected cashflows over an assumed ten-year horizon. It should be noted that the assumed investment horizon could

Rank	Country/territory	Region		Overall Score	Investment Performance	Market Fundamentals	Listed Vehicles	Regulatory & Legal	Transaction Processes	Sustainability
Rank out of 99 territories	...m	EUR		1.25	1.02	1.77	1.00	1.15	1.00	1.80
2	United States	AM		1.34	1.12	1.48	1.00	1.38	1.45	1.70
3	France	EUR		1.34	1.25	1.77	1.31	1.21	1.00	1.70
4	Australia	AP		1.38	1.18	1.64	1.00	1.41	1.13	2.10
5	Canada	AM		1.44	1.49	1.79	1.17	1.23	1.20	1.90
6	Netherlands	EUR		1.54	1.60	1.64	1.40	1.31	1.15	2.50
7	Ireland	EUR		1.69	2.25	1.93	1.00	1.26	1.00	2.60
8	Sweden	EUR		1.76	1.81	2.61	1.32	1.25	1.30	2.50
9	Germany	EUR		1.76	1.94	2.05	1.39	1.49	1.30	2.57
10	New Zealand	AP		1.77	1.83	1.96	1.01	1.59	1.00	3.63
11	Belgium	EUR		1.84	2.10	2.29	1.10	1.76	1.10	2.50
12	Japan	AP		1.88	1.60	2.67	2.05	1.47	1.80	2.20

Figure 5.3 JLL 2022 Transparency index scores for "highly transparent" markets.

Source: Jones Lang LaSalle.

potentially have a significant bearing on the calculated appraisal figure, and the anticipated holding timeframes for an investment should likely be the adopted assumption by investors.

The assumed exit value is calculated using a US convention cap rate of 5.5% and sales costs of 2% are also incorporated. Making use of the commercial real estate yield or cap rate formula shown in Section 4.2.2 and a modelled rental growth assumption of 2% p.a., this cap rate includes a deprecation or obsolescence provision of 0.50% p.a. as follows:

5.50% exit cap rate in year 10:
+ 4.00% Risk-free rate
+ 3.00% Risk premium
– 2.00% Rental growth
+ 0.50% Depreciation
= 5.50% Exit cap rate

In this instance, the current value of the 10-year projected stream of income and capital returns was 25.13 million, which is the estimated investment worth. Assuming the same set of cashflow projections, if an investor were able to purchase this asset below this value, a higher return than 7% would be achievable. Thus, an investor would be projected to make a total return above that required, pointing to relative value being available. Of course, the opposite would be true if pricing were above this appraised level.

5.4.1 Valuation sensitivity analysis

In a similar vein to the projected performance sensitivity analysis provided in Section 4.4.1, a valuation's sensitivity to changes in underlying assumptions can be assessed. This is most helpful to investors judging the worth of an asset under a range of scenarios, whether these be driven by market variables or asset-specific considerations. Whilst independent valuers are required to provide a single "point estimate", investors may undertake sensitivity analysis if they believe valuers will have to make significant adjustments to the underlying assumptions due to a view on anticipated market conditions or advanced knowledge of a significant asset-specific issue such as a material lease event.

Figure 5.5 provides an example of such analysis and highlights the variation in the calculated valuation across a range of market rental growth and discount rate assumptions. The central projections from Figure 5.4 are highlighted. In this example, the exit cap rate underlying the valuation is held constant at 5.5% instead of varying per the above-worked formula as outlined in Section 5.3.3. This example highlights how an investor's estimate of appraised worth can shift in response to a range of key assumptions and projections.

5.4.2 Incorporating sustainability considerations into valuations

Sustainability and impact considerations have a significant bearing on commercial real estate markets and as such now need to be incorporated into appraisals. By way of example, current European Union regulations render buildings with a low Energy Performance Certificate ("EPC") rating (F and G ratings) obsolete by 2030, with certain governments taking a more stringent approach. For example, in the Netherlands, office buildings with an EPC rating below C will no longer be permitted for that functional use,[2] with an A rating to be imposed by 2030. There is also increasing evidence of more sustainable energy buildings

Quarter End Date	Projected Cashflows	Present Value of a 7% Discount Rate	Present Value of Cashflows	Valuation / Appraisal
31-Mar-22				25,135,899
30-Jun-22	361,250	0.983	355,212	
30-Sep-22	361,250	0.967	349,209	
31-Dec-22	361,250	0.950	343,308	
31-Mar-23	121,075	0.935	113,159	
30-Jun-23	366,350	0.919	336,676	
30-Sep-23	366,350	0.903	330,987	
31-Dec-23	366,350	0.888	325,394	
31-Mar-24	-67,780	0.873	-59,196	
30-Jun-24	180,353	0.859	154,880	
30-Sep-24	180,353	0.844	152,263	
31-Dec-24	193,084	0.830	160,256	
31-Mar-25	84,686	0.816	69,126	
30-Jun-25	185,764	0.803	149,097	
30-Sep-25	185,764	0.789	146,577	
31-Dec-25	185,764	0.776	144,101	
31-Mar-26	312,734	0.763	238,583	
30-Jun-26	116,843	0.750	87,649	
30-Sep-26	416,843	0.737	307,408	
31-Dec-26	416,843	0.725	302,214	
↓	↓	↓	↓	
30-Jun-31	479,427	0.535	256,429	
30-Sep-31	479,427	0.526	252,096	
31-Dec-31	479,427	0.517	247,837	
31-Mar-32	32,884,066	0.508	16,715,043	

Figure 5.4 Projected unlevered quarterly cashflows and valuation −7% discount rate assumed.
Source: The authors.

Sensitivity
Table 1

		Market Rental Growth				
		-2.0%	0.0%	2.0%	4.0%	6.0%
	9.0%	18.140	19.777	21.587	23.589	25.801
	8.0%	19.511	21.299	23.277	25.464	27.881
Discount Rate	7.0%	21.018	22.973	25.136	27.528	30.172
	6.0%	22.677	24.817	27.184	29.802	32.697
	5.0%	24.505	26.849	29.442	32.311	35.484

Figure 5.5 Valuation sensitivity analysis (millions) assuming a 5.5% exit cap rate.
Source: The authors.

garnering economic benefits. For instance, Fuerst and McAllister (2011) found that certified green commercial properties demonstrated resale prices between 10% to 31% above their non-certified peers.

This has been recognised by industry bodies, with, for example, RICS (2022) guidance stating the following:

> *Valuers should ensure that, as far as reasonably possible, up-to-date information on environmental and physical risks is gathered in respect of the subject property and considered when comparing it to others used as part of the evidence base.*

The question then remains how to reflect sustainability consideration in appraisals. Sayce et al. (2022) outline the factors that valuers should consider and whether they impact projected income and expenditures or discount rates through the risk premium applied. The requisite capital expenditure provisioning to appropriately upgrade properties to ensure regulatory compliance is a clear near-term issue that asset owners and valuers need to address.

One could look at this being a typical depreciation consideration which ultimately deteriorates rental value as other considerations have over time. However, the scale of the sustainability issue is hugely significant and, as highlighted above, could lead to heightened rates of depreciation to the point of functional obsolescence. Increased risk premiums for less sustainable assets could reflect greater expected income volatility or reduced liquidity, as occupier and investor preferences increasingly focus on this topic. Regulations could also evolve further and become even more punitive for unsustainable assets on a global basis, increasing the need to reflect them in appraisals.

Quiz questions

1 What are the differences between the three market valuation approaches described in this chapter?
2 What factors can drive differences in risk premium and discount rates across markets?
3 How can sustainability considerations be fed into DCF-based valuation models?
4 Using the illustrative DCF model, what would be the change in the calculated investment worth valuation adjustment if budgeted capital expenditures doubled?
5 Using the illustrative DCF model illustration example of cash flow in Figure 5.3, fill in the following sensitivity analysis matrix for the investment worth calculated across a range of CPI growth and exit (year 10) cap rate assumptions.

		CPI			
	0.0%	*1.0%*	*3.0%*	*5.0%*	*7.0%*
Exit Cap Rate	**6.5%**				
	6.0%				
	5.5%		25.136		
	5.0%				
	4.5%				

Discussion questions

- Why do discount rates vary by sector and by country?
- How would you estimate discount rates for non-traditional commercial real estate sectors?
- How can you quantify asset-specific risk premiums using subjective and objective frameworks?

Research/Dissertation topics

- An assessment of the strengths and weaknesses of valuation approaches across jurisdictions.
- How can discounted cashflow assessments of worth be "greened"? *This would be well suited to case studies.*

Notes

1 See www.jll.com for the latest version.
2 https://business.gov.nl/regulation/energy-labels/

References

Addae-Dapaah, K., Ho, D. & Glascock, J.L. (2014). International direct real estate risk premiums in a multi-factor estimation model. *Journal of Real Estate Finance and Economics*, 51 (1): 52–85.

Breidenbach, M., Mueller, G. & Schulte, K.W. (2006). Determining real estate betas for markets and property types to set better hurdle rates. *Journal of Real Estate Portfolio Management*, 12 (1): 73–80.

Crosby, N., Hutchison, N., Lusht, K. & Yu, S.M. (2018). Valuations and their importance for real estate investments. In: MacGregor, B. D., Schulz, R. and Green, R. K. (eds.). *Routledge companion to real estate investment*. Routledge, Abingdon.

Fuerst, F. & McAllister, P. (2011). Green noise or green value? Measuring the effects of environmental certification on office values. *Real Estate Economics*, 39 (1): 45–69.

RICS. (2021). *Sustainability and ESG in commercial property valuation and strategic advice, Global*, 3rd edition. Royal Institution of Chartered Surveyors. Please see the following: https://www.rics.org/profession-standards/rics-standards-and-guidance/sector-standards/valuation-standards/sustainability-and-commercial-property-valuation

Sayce, S., Clayton, J., Devaney, S. & Van der Wetering, J. (2022). Climate risks and their implications for commercial property valuations. *Journal of Property Investment and Finance*, 40 (4): 430–443.

Private markets

6 Private real estate fund structures

Objectives

At the end of this chapter, you will understand:

- The risk profile of the different private real estate fund (PREF) strategies.
- The development of the PREF market universe.
- Different PREF vehicle structures.
- The impact of PREF fee economics.

Key concepts

- Risk and Return Style guidelines of PREFs to help investors allocate correctly according to their risk budget.
- Drivers behind the growth of the PREF universe.
- How the structure of a PREF affects investor returns.
- Performance fees for the fund manager and their calculation.

6.1 Introduction to private real estate funds

The UK-based Association of Real Estate Funds (AREF) defines PREFs as follows:

> *A property fund is a collective investment scheme with a portfolio comprising mainly of direct property but may also include other property-related interests. Property funds take a number of different legal structures depending on their domicile and target customer.*

A PREF is a non-traded investment vehicle in which investors co-mingle their capital with the aim of providing direct private real estate total returns. PREFs have a dedicated fund manager who is responsible for all investment and operational activities, including the execution of the stated strategy and implementing the day-to-day decision-making. Within the fund structure, investors pool capital to create a larger supply of capital than could be achieved individually. This enables access to a wider selection of assets and exposure to potential real estate investments than would otherwise have been possible individually.

Commercial real estate as an asset class has seen strong, and growing investor interest in recent decades. However, commercial real estate assets are naturally heterogeneous, and a typical barrier to direct investment for investors is their illiquidity and "lumpy" lot sizes. Consequently, constructing well-diversified direct real estate portfolios is much

DOI: 10.1201/9781003298564-8

more difficult than for more traditional liquid asset classes, such as equities and bonds. It also requires significant capital and resources to acquire and manage the assets. By way of example, Callender et al. (2007) found that large commercial real estate portfolios were required to track the UK commercial real estate market. Indeed, to reduce tracking error to 2%, their study found that approximately 60 assets were required, which would mean a capital outlay of approximately £800 million.

In a similar vein, the biggest barrier to cross-border direct real estate investment is scale. Even the largest of institutional investors are not able to construct portfolios of a satisfactorily diversified size. By way of example, Baum and Kennedy (2012) showed that approximately 500 assets would be required to achieve a tracking error of 4% against a global direct property benchmark. The authors note that constructing a portfolio of 300–500 assets directly assuming 50% leverage and a 10% allocation to real estate as an asset class would imply a portfolio ranging between $30 billion USD and $125 billion USD, a figure well out of reach of all but the largest institutional investors. However, exposures of this scale can be accessed using portfolios of PREFs.

Finally, other key rationales for investments in PREFs are investors' internal resource constraints and their available capital to deploy. An investor's level of internal real estate team resources will naturally be a function of their size. Certain more operationally intensive sectors require specialist management expertise that is not widely available. Even the largest institutional investors use PREFs to facilitate at least their own non-domestic real estate allocations and access specialist managers with capabilities in a sector vertically.

6.2 Defining the private real estate fund market universe

PREFs execute either balanced or specialist strategies. Balanced strategies provide investors with a well-diversified market exposure either at a country or region level across multiple property types. Conversely, specialist funds focus on a particular market segment or niche, for example, an Australian industrial fund. There is also a range of risk profiles available to investors. The industry has adopted what are termed "fund styles" as risk classification categories. To illustrate, the INREV guidelines for assigning PREFs to a fund style are shown in Figure 6.1.

The INREV guidelines attribute risk at a direct property level as being attributable to the degree of market or operational risk taken in the form of leasing/vacancy and/or

	Core ≤ 40% LTV	Core ≥ 40% LTV	Value Add	Opportunity
Total % of non-income-producing investments	≤ 15%		> 15% -≤ 40%	> 40%
Total % of (re)development exposure	≤ 5%		> 5% -≤ 25%	> 25%
% Of total return derived from income	≥ 60%			
Maximum LTV	≤ 40%	> 40%	> 40% -≤ 60%	> 60%

Figure 6.1 INREV real estate fund style classification criteria.

Source: INREV, 2012.

Study	Market Beta Estimates
Fuerst and Matysiak (2012)	Market betas of 1.1 to 1.2 for Value Add funds and 1.3 for Opportunistic funds relative to direct real estate performance
Alcock et al. (2013)	Market betas of 1.1 to 1.2 for Value-Add funds and 1.7 to 1.8 for Opportunistic funds relative to direct real estate performance
Pedersen and Page (2014)	Market beta of 0.5 to 0.8 depending upon PREF risk profile relative to general equities performance
Ang et al. (2018)	Market beta of 0.8 for a sample of closed-end PREFs relative to publicly traded real estate performance
Farrelly and Stevenson (2019)	Market beta of 2.0 to 3.3 depending upon PREF risk profile relative to direct estate
	Market beta of 0.8 to 1.0 depending upon PREF risk profile relative to publicly traded real estate performance.
	Market beta of 0.6 to 0.8 depending upon PREF risk profile relative to general equities performance

Figure 6.2 Select studies quantifying the market betas of private real estate funds.

Source: Authors' summary.

development risk. It is worth noting that these do not address market risk in terms of geography or sector. Financial leverage is then the other source of key risk with the LTV thresholds defined, and a 40% LTV delineating the lower-risk core fund style categories. More broadly, the industry refers to Core vehicles utilising financial leverage above a 40% LTV as Core-Plus.

Value-added funds hold portfolios where the underlying real estate assets include a mix of income-producing investments and assets carrying operational risk, for example, in the form of vacancy and (re)development. Opportunistic funds are more growth-orientated with most of their target performance expected in the form of appreciation. Thus, they should exhibit higher performance volatility. This may be due to a variety of characteristics, such as exposure to development, significant leasing/occupancy risk, elevated levels of financial leverage, or a combination of risk factors.

However, whilst these INREV criteria and others like them are helpful to expediently classify a PREF's risk profile, they remain descriptive and do not provide objective estimates of PREF risk profiles. A relatively small number of studies have sought to do this by using various regression-based methodologies centred on the capital asset pricing and factor models to assess the incremental risk profiles of PREFs relative to direct or public real estate and general equity market exposures. These estimates are summarised in Figure 6.2. As one would expect, these estimates reflect meaningful incremental risk above direct unleveraged performance and similar levels of implied risk to publicly traded securities.

6.3 Growth of the PREF market universe

The growth seen in the private real estate fund market has helped facilitate both domestic investors and growing cross-border investment into the real estate asset class across the world. The universe of private real estate funds has grown dramatically, truly developing in the pre-GFC era. Recent global capital-raising trends are shown in Figure 6.3. Whilst

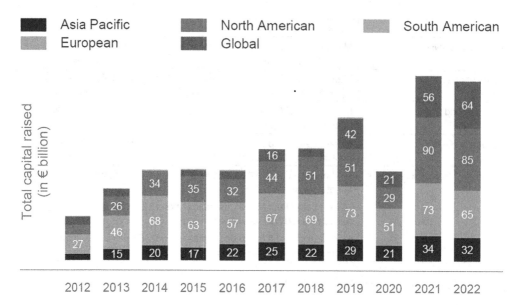

Figure 6.3 Aggregate private real estate fund capital raised – EUR billion.

Sources: ANREV, INREV, NCREIF, 2023.

there is an apparent cyclical dimension to the capital raised, since 2012 there has been steady growth in capital raised for these investment conduits coming out of the GFC, with a recent slowdown in 2020 being due to the impact of the COVID-19 pandemic.

6.4 Fund structures

Another distinguishing feature of a PREF is its lifespan, with funds having either open-ended or closed-ended structures. Open-ended funds are perpetual vehicles and can raise capital as and when investor demand dictates. Units in open-ended funds are issued and cancelled in response to investor demand, and, importantly, investors have the right, subject to predetermined redemption procedures, to redeem their holdings. To protect the interests of ongoing investors, redemptions requiring the sale of properties in adverse market conditions may be subject to deferral. As such, this liquidity activity may have an impact on the performance delivered by open-ended real estate funds. These fund structures are common for lower risk-return Core or Core-Plus strategies, although there is a wide range of US Value Add PREFs available in an open-ended format.

In contrast, closed-ended funds have a fixed number of shareholdings or partnership interests and have a finite life with a set maturity date, commonly from six to seven years and up to twelve years. Figure 6.4 shows a schematic of a closed-ended private real estate fund structure, in this case, a limited partnership (a legal structure often used for closed-ended PREFs). The private real estate fund structure or limited partnership is the legally formed conduit that receives investor capital commitments and then makes investments on behalf of the investors.

Investors will typically provide the vast majority of the fund's capital alongside co-investment from the manager. According to Preqin[1] US-focused Value Added and

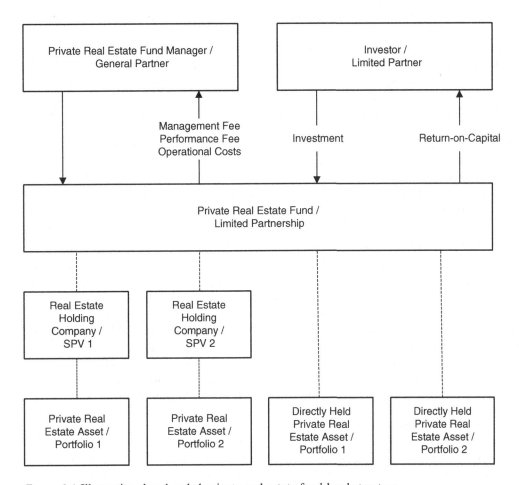

Figure 6.4 Illustrative closed-ended private real estate fund legal structure.

Source: The authors.

Opportunistic funds, the average level of co-investment by the manager has ranged from 2% to 4.5% of total fund commitments. The capital commitments made cannot be rescinded during the life of the fund, although a fund's legal provisions will often dictate the period it is able to deploy commitments – this is often known as the "investment period". After this period, no commitments to new investments can be made by the fund. The "Manager" or "General Partner" is an entity established and owned by a specific private real estate management firm.

PREFs are governed by a set of legal documents and contracts that dictate the investment strategy, fee arrangements, and duration, among other governance aspects. For example, a fund will typically have a set of binding investment restrictions to ensure a certain risk profile and permissible scope of strategy. Typical parameters include deal concentration, market exposure, financial leverage (deal and PREF levels) and development activity exposure limits. PREFs will also often have investor adviser boards, whereby the largest investors in a PREF have additional oversight and voting rights concerning the activities of a PREF, sometimes with independent director support.

Whilst there are a range of commercial and legal considerations for investors, one often contentious governance issue of note is that of manager removal rights. Typically, PREFs provide for "non-cause" and for "cause removal", both of which require majority or usually super majority votes from the entire investor base. For-cause removal is where the manager has committed wrongful acts such as fraud, wilful misconduct, or gross negligence or has gone bankrupt. Even in these circumstances, they may still have rights to performance fee economics, and this may have negative governance or reputational ramifications for the investors and their stakeholders. Non-cause removal occurs in other circumstances where the investors wish to replace the manager. However, in these instances, the manager will continue to receive performance fees, making this event costly for the investors.

6.5   Private real estate fund economics

Managers of PREFs will receive recurring investment management fees and, in many instances, performance-based fees. The recurring fee element can be based upon total commitments, drawn capital (at cost) or the fund's most recent valuation – either the NAV or possibly the gross value of the real estate assets held. Open-ended structures tend to levy management fees upon valuations whereas closed-ended PREFs executing higher risk-return strategies charge fees on committed capital (during the investment period) and invested capital. Typical management fees for closed-ended higher risk-return strategies are 1.5% p.a. of commitments during the investment period and 1.5% p.a. of drawn (cost) capital thereafter. The fees on commitments during the investment period are for the manager at the early stages of a PREF's lifecycle as the portfolio is being constructed.

Performance fees can be payable on both an absolute and relative total return basis. These seek to reward good manager performance and incentivise management teams. Relative return-based performance fees generally only apply to lower risk-return, open-ended PREFs and are typically calculated on a rolling basis, for example, over three-year periods. Absolute return-based fees apply to nominal return targets and the manager receives a stated percentage of the profit delivered ("carried interest") on invested capital above a preferred return level to investors. These are mostly applied to closed-ended PREFs, with calculations applied over the full lifecycle of the fund, but they are increasingly becoming commonplace for perpetual PREFs, with calculations made over rolling annual periods. The typical preferred return is an IRR objective. By way of example, Van der Spek (2017) found that for a sample of 324 Value Added and Opportunistic funds, the average preferred return was approximately 10% and the carried interest, 20%. In recent years, preferred return levels of 8% and 9% have been more commonplace.

These absolute return-based remuneration structures can have several features that will have a material impact on net-of-fee performance to investors. The inclusion of a catch-up mechanism is especially impactful, and this provision provides enhanced profit sharing for the manager. Once investors have received their capital invested and preferred return on that invested capital, the manager "catches up" to the investors' profit level based on a defined percentage split, which is typically 50%–100%. Then a defined lower profit split applies to all subsequent profits.

6.5.1 *Private real estate fund cashflow waterfalls*

Before turning to the net economic impact of performance fee structures, PREF cash flows are distributed according to the following waterfall:

1 The PREF pays all recurring fees and costs including management fees.
2 The investors receive their invested capital.
3 The investors receive the preferred return on their invested capital.
4 If there is a catch-up provision, the manager will receive an enhanced share of proceeds until they have caught up to the defined percentage profit share, typically 20%.
5 The manager is paid "carried interest", whereby they receive a fixed minority share of the remaining distributions, typically 20%.

The timing of these performance fee payments can also vary with what is termed "European" waterfalls, where the manager does not receive any performance fee payments until the investors have received all their invested capital and the preferred return on this. The "American" waterfall model is favourable to managers, enabling them to be paid on a deal-by-deal basis, meaning investors will not have received their invested capital when there are profitable earlier-stage realisations in the PREF's lifecycle. In both instances, performance fees can be payable before the end of the PREF's lifecycle. To protect investors, there are clawback and holdback provisions.

Clawback provisions enable investors to seek returns of performance fee payments when subsequent performance has deteriorated such that managers were overpaid through earlier carried interest payments. Holdbacks help protect against the credit risk aspect of this, as the managers need to be able to pay back, and so they require managers to hold back a percentage of the performance fee paid, often 50%, in escrow accounts, which can only be released upon the maturity of the PREF.

To illustrate the workings and dynamics of PREF fee economics, hypothetical cashflows and a gross-to-net return dilution analysis are shown in Figures 6.5 and 6.6.

Figure 6.5 demonstrates how the carried interest is calculated for a PREF with a 9% preferred return and 20% carried interest. As can be seen, Row D calculates the preferred return level due to the investor and this is deducted from Row C. The carried interest share is calculated using this residual value. The impact of a 50% catch-up provision is such that once the investor has received their invested capital and preferred return on that, the manager receives a 50% share of the profit generated thereafter until they have received 20% of total profits. As can be seen in Figure 6.6, the catch-up provision materially dilutes investor returns between approximately 11% and 15% of the gross IRR level, after which the manager is caught up and the gross-to-net spread levels move in a broadly parallel fashion. To provide further context, Baum and Farrelly (2009) give the performance dilution from fee structures for a range of funds managed by a particular manager, shown in Figure 6.7.

The performance fee structures described above seek to incentivise and align managers to the success of the fund, but instead could simply reward risk-taking in the form of execution risk at the asset level and using excess financial leverage rather than outperformance in the form of true "alpha". In the event performance is negative, investors participate fully in the loss but conversely their upside can be significantly curtailed depending upon the performance fee structure of a PREF. The impacts of these empirical figures and illustrative modelling exercises highlight that investors need to be

ID	Year	Calculation	IRR	1	2	3	4	5	6	7	8
A	Gross Investor Cashflow		14.0%	-30.0	-30.0	-28.7	-5.9	31.7	70.3	53.9	6.4
B	Annual Management Fees			0.8	1.5	1.5	1.5	1.2	0.7	0.2	0.0
C	Pre-Performance Fee Cashflow	A - B	12.2%	-30.8	-31.5	-30.2	-7.4	30.5	69.6	53.7	6.4
D	Preferred Return		9.0%	-30.8	-31.5	-30.2	-7.4	30.5	69.6	40.2	0.0
E	Excess Profit	C - D								13.6	6.4
F	Carried Interest	E * 20%								2.7	1.3
	Net Investors Cashflow	C - F	11.6%	-30.8	-31.5	-30.2	-7.4	30.5	69.6	51.0	5.1
	Total Gross Profit	Sum (A)	67.7								
	Total Net Profit	Sum (G)	56.4								
	Fee Dilution	1 - (G / A)	-16.7%								

Figure 6.5 Illustrative PREF cashflow for an assumed $100 commitment 1.5% p.a. management fee on committed and invested capital and a 20% carried interest over a 9% preferred return performance fee with no catch-up.

Source: The authors.

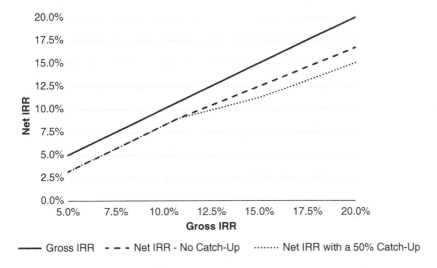

Figure 6.6 Illustrative PREF gross-to-net IRR dilution, excluding and including a 50% catch-up provision.

Source: The authors.

Fund	Gross IRR	Net IRR	Fee impact	Fee impact %
1	29.0%	25.0%	4.0%	13.8%
2	17.0%	13.0%	4.0%	23.5%
3	33.0%	25.0%	8.0%	24.2%
4	35.0%	30.0%	5.0%	14.3%
5	27.0%	21.0%	6.0%	22.2%
6	46.0%	37.0%	9.0%	19.6%
7	21.0%	16.0%	5.0%	23.8%
8	34.0%	27.0%	7.0%	20.6%
9	16.0%	13.0%	3.0%	18.8%
10	20.0%	15.0%	5.0%	25.0%
11	18.0%	14.0%	4.0%	22.2%
12	20.0%	16.0%	4.0%	20.0%
13	14.0%	12.0%	2.0%	14.3%
14	20.0%	15.0%	5.0%	25.0%
Mean	25.0%	19.9%	5.1%	20.5%

Figure 6.7 Case study performance dilution from fee structures for a single manager.

Source: Baum and Farrelly (2009).

aware of the performance fee impacts of the waterfall structure in the PREF under consideration. See also van der Spek (2017) for a review of PREF fee structures and Metrick and Yasuda (2009) for a thorough analysis of the management fee arrangements in private equity funds that are structured in an equivalent manner to closed-ended private real estate funds.

6.5.2   *Tax considerations*

There is a myriad of available investment structures available to select for PREFs. These have varying degrees of tax efficiency, which will vary depending on an underlying investor's jurisdiction and status. In most countries, there are local structures available that provide efficient access for local tax-exempt investors, particularly pension funds. These may provide efficient access or, indeed, may not permit other local or non-domestic investors who do not meet specific criteria. They are typically perpetual vehicles. For example, private REITs in the US mirror their listed counterparts in that they provide tax-efficient access to this market for institutional investors. Income distributions from these will be non-taxable for US institutions but admissible foreign investors may incur withholding and other income taxes locally as appropriate. However, for all investors, the sale of shares in the private REIT does not trigger local US capital gains taxation whereas the sale of a direct asset does, which is a key advantage of this structure, especially for foreign investors.

Closed-ended PREFs such as limited partnerships are typically "tax transparent" wrappers in that they pass through all distributions (income and capital) from the underlying investments, with no tax levied at the fund level itself. Instead, underlying investments incur tax incidence locally and investors are taxable at their entity level depending upon their respective jurisdiction and status.

Quiz questions

1 What are the four styles of a private real estate fund?
2 What are the key differences between Core and Core-Plus Funds?
3 Identify the roles of a general partner and a limited partner.
4 What metrics do you require to determine an investor's net cash flow in a PREF?
5 What is the typical lifespan of a closed-end fund?

Discussion questions

• What are the advantages of a PREF relative to a public real estate vehicle (REIT)?
• What are the advantages of closed-ended funds vs. open-ended funds from a fund manager's viewpoint?

Research/Dissertation topics

• What are the key drivers of performance for a PREF and how would you measure them?
• A discussion of alignments of interest and potential adverse agency effects arising from a range of PREF fee economics arrangements.

Note

1 Preqin's average benchmark fund terms for a $1 billion equity fund. At www.Preqin.com

References

Alcock, J., Baum, A., Colley, N. & Steiner, E. (2013). The role of leverage in the performance of private equity real estate funds. *Journal of Portfolio Management*, 39: 99–110.

Ang, A., Chen, B., Goetzmann, W.N. & Phalippou, L. (2018). Estimating private equity returns from limited partner cash flows. *Journal of Finance*, 73: 1751–1783.

Baum, A. & Farrelly, K. (2009). Sources of alpha and beta in property funds: A case study. *Journal of European Real Estate Research*, 3: 218–234.

Baum, A. & Kennedy, P. (2012). Aligning asset allocation and real estate investment: some lessons from the last cycle. Working Papers in Real Estate & Planning, University of Reading, 07/12.

Callender, M., Devaney, S., Key, T. & Sheahan, A. (2007). Risk reduction and diversification in UK commercial property portfolios. *Journal of Property Research*, 24: 355–375.

Farrelly, K. & Stevenson, S. (2019). The risk and return of private equity real estate funds, *Global Finance Journal*, 42: 1–13.

Fuerst, F. & Matysiak, G.A. (2012). Analyzing the performance of non-listed real estate funds: A panel data analysis. *Applied Economics*, 45: 1777–1788.

INREV (2012). *INREV Style Classification*, Revised Edition. Amsterdam, February 2012.

Metrick, A. & Yasuda, A. (2009). The economics of private equity funds. *Review of Financial Studies*, 23: 2303–2341.

Van der Spek, M. (2017). Fee structures in private equity real estate. *Journal of Real Estate Research*, 39: 319–348.

7 Valuing private real estate vehicles

Objectives

At the end of this chapter, you will understand:

- The private real estate vehicles available to investors.
- How discount rates can be adjusted to match the profile of a levered real estate investment.
- The valuation of PREFs and considerations for their structure.
- Secondary transactions and how performance projections and valuations can be undertaken.

Key concepts

- Levered real estate transaction valuations.
- Open vs. closed-ended PREF considerations.
- PREF accounting practices.
- Valuations of PREFs.
- Secondary markets and transactions.

7.1 Introduction

Direct real estate assets and portfolios can be held by investors within a number of private vehicles across jurisdictions including both onshore, that is, vehicles incorporated/registered in the country where assets are located, and offshore entities. Offshore entities typically exist to enable a broad range of investors to access real estate on an efficient basis, with materially reduced levels of tax incidence providing the impetus for investors to hold their assets in these locations. These vehicles have played a critical role in the development of the global real estate capital market. They have been instrumental in facilitating cross-border investments as real estate investors have broadened their opportunity set beyond domestic markets to benefit from diversification and a broader range of investment options.

Chapter 12 provides an overview of PREFs, which have been key structures for all institutional investors. But there also exists a range of corporate entities, many of which are used by PREFs for their underlying holdings. These corporate entities can take the form of what are often termed real estate operating companies (REOCs), which can be internally managed with dedicated staff or be externally managed in a similar vein as a PREF fund, as described in Section 12.1.4. These tend to hold sizeable

DOI: 10.1201/9781003298564-9

portfolios, with investor governance matters controlled via shareholder and potentially board representation.

Individual assets and portfolios can be held in special-purpose vehicles and corporate entities to facilitate their efficient holding for tax and other considerations. These entities also enable investors to ringfence any potential liabilities arising from the assets, for example, issues around debt financing, insurance matters or other claims. Both PREFs and REOCs often hold their real estate assets in such vehicles, with all financial reporting consolidated on a look-through basis.

7.1.1 Private real estate structures

7.1.1.1 Adjusting direct real estate risk premium for levered structures

When utilising financial leverage for real estate investments, the risk-return profile will change significantly. Thus, for the valuation of private real estate vehicles such as corporate entities, valuation methodologies that account for this need to be employed. To extend the overview provided in Section 5.3.2 for estimating direct (unleveraged) real estate discount rates, levered investors can adjust unleveraged direct real estate betas using the following formula:

$$Levered\ Beta = Unlevered\ Beta * (1 + (1 - Tax\ Rate) * Debt/Equity))$$

Figure 7.1 shows the estimated levered discount rates calculated using this formula when applying it to the unleveraged discount rate assumptions underlying Figure 5.3 of a 4%

Asset/Portfolio Beta	1.0
Tax Rate	15%
Risk Free Rate	4.0%
Risk Premium	3.0%

LTV	Levered Beta	Risk Premium	Discount Rate
0%	1.00	3.0%	7.0%
5%	1.04	3.1%	7.1%
10%	1.09	3.3%	7.3%
15%	1.15	3.5%	7.5%
20%	1.21	3.6%	7.6%
25%	1.28	3.9%	7.9%
30%	1.36	4.1%	8.1%
35%	1.46	4.4%	8.4%
40%	1.57	4.7%	8.7%
45%	1.70	5.1%	9.1%
50%	1.85	5.6%	9.6%
55%	2.04	6.1%	10.1%
60%	2.28	6.8%	10.8%
65%	2.58	7.7%	11.7%
70%	2.98	9.0%	13.0%
75%	3.55	10.7%	14.7%

Figure 7.1 Adjusted discount rates for the level of financial leverage used.

Source: Authors' calculations.

risk-free rate and 3% risk premium alongside. Figure 7.1 also assumes an asset/portfolio beta of 1.0x and effective tax rate of 15%, which results a broad range of estimated discount rates or required returns for varying LTV levels. The calculated levered beta is applied to the 3% risk premium assumption to, in turn, provide a discount rate estimate.

This methodology provides a simple framework for developing more appropriate discount rates for the valuation of private real estate vehicles. Investors may wish to consider a range of other factors relating to the holding structure. These may include liquidity profile, ease of execution and administration, and the potential impacts of changing regulations and laws that could affect the attraction or viability of using it for an investor. By way of example, Farrelly and Stevenson (2019) found in their asset pricing study of higher risk-return closed-ended PREFs that an approximate 2% p.a. liquidity premium should be applied, which was consistent with findings from private equity studies that a positive premium should apply to account for the relative illiquidity of these entities.

7.1.2   *Putting it all together – a DCF model valuation model for a levered structure*

In this section, the worked example for an annual direct property cashflow model from Chapter 5 is extended to include a DCF model for an interest in an illustrative levered private real estate structure, in this case, a corporate entity. This is shown in Figure 7.2, where the annual direct real estate cash flows from Figure 4.2 also have the impact of corporate entity costs (row 6), the use of debt finance (rows 10 to 12), and investor level tax (rows 14 to 18) included. See the accompanying Excel file for this chapter, where the full set of assumptions underlying these projections can be reviewed.

Corporate entity costs include set-up and ongoing fees, which are assumed to inflate at 3% per annum after the first year. For more substantive private real estate companies, further costs comprise general and administrative expenses, including staff costs and business operations overheads. A relatively simple 50% LTV debt finance package is assumed with a 1% arrangement fee and a fixed 4% annual interest rate cost. Income and capital gains taxation are both assumed to be at a 15% rate, with interest costs being tax deductible; that is, income net of interest costs is taxed. Capital gains tax is levied on the total capital outlay, with all purchase and sales set-up and capital expenditures included in the aggregate cost basis (row 16), from which projected taxable gains are calculated.

The appraisal or valuation of the equity interest in this vehicle is estimated by calculating the current value using a levered discount rate as the basis for a DCF valuation. The levered discount rate calculated is as per Figure 7.1 for the 50% LTV estimate at 9.6%. This is used to calculate the current value of projected net of tax cashflows over an assumed five-year horizon, when it is assumed, the investment will be exited. In this instance, the current value of the five-year projected stream of income and capital returns was 12.04 million, which is the estimated investment worth of this entity for an equity investor.

7.1.2.1   *Valuation sensitivity analysis (millions)*

In a similar vein to Figure 5.5, Figure 7.3 shows the illustrative private real estate vehicle's equity valuation sensitivity to changes in underlying assumptions. In this example, we do not focus on the underlying direct real estate performance drivers (e.g., rental growth, exit yield), rather the assigned unleveraged beta and investor effective tax rate are sensitised. The central projection from Figure 7.2 is highlighted. This example highlights how an investor's estimate of appraised worth can shift in response to a range of key assumptions and projections relating to views on perceived risk profile and the level of tax incidence

Input assumptions

Entry Cap Rate	6.00%
Exit Cap Rate	5.50%
Purchase Price	22,835,417
Acquisition Costs	4.00%
Sales Costs	2.00%
Loan-to-Value Employed	50.00%
Loan Balance	11,417,708
All-In Cost (Interest Rate plus Margin)	4.00%
Loan Arrangement Fees	1.00%
Corporate Entity Setup Cost	50,000
Recurring Corporate Entity Costs	30,000
Cost Inflation	3.00%
Income Tax	15.00%
Capital Gains Tax	15.00%
Asset/Portfolio Beta	1.0
Tax Rate	15%
Risk Free Rate	4.0%
Risk Premium	3.0%
Levered Discount Rate	9.6%

Projected cashflows

Item / Variable	Row	0 Current Mar-22	1 Year 1 Mar-23	2 Year 2 Mar-24	3 Year 3 Mar-25	4 Year 4 Mar-26	5 Year 5 Mar-27	6 Year 6 Mar-28
Property Net Operating Income	1		1,370,125	1,388,279	641,977	661,236	1,583,100	1,603,533
Purchase Price	2	(22,835,417)						
Transaction Costs	3	(913,417)						
Leasing Fees	4						(106,121)	
Capital Expenditure	5	(50,000)	(150,000)	(150,000)	0	(300,000)	(50,000)	
Corporate Entity Costs	6		(30,000)	(30,900)	(31,827)	(32,782)	(33,765)	
Projected Exit Price	7						29,155,137	
Sales Costs	8						(583,103)	
Unlevered Cashflows	9	**(23,798,833)**	**1,190,125**	**1,207,379**	**610,150**	**328,455**	**29,965,248**	
IRR		7.5%						
Profit-On-Cost		1.40x						
Loan Balance	10	(11,417,708)	(11,417,708)	(11,417,708)	(11,417,708)	(11,417,708)	(11,417,708)	
Arrangement Fees	11	(114,177)						
Interest Costs	12		(456,708)	(456,708)	(456,708)	(456,708)	(456,708)	
Levered Gross of Tax Cashflows	13	**(12,266,948)**	**733,417**	**750,670**	**153,442**	**(128,254)**	**18,090,832**	
IRR		10.5%						
Profit-On-Cost		1.59x						
Pre-Tax Income	14		883,417	900,670	153,442	171,746	986,506	
Income Tax Payable	15		(132,513)	(135,101)	(23,016)	(25,762)	(147,976)	
Capital Cost	16	(23,913,010)	(24,063,010)	(24,213,010)	(24,213,010)	(24,513,010)	(24,563,010)	
Taxable Capital Gain	17						4,009,024	
Capital Gains Tax Payable	18						(601,354)	
Levered Net of Tax Cashflows	19	**(12,266,948)**	**600,904**	**615,570**	**130,426**	**(154,016)**	**17,341,502**	
IRR		9.1%						
Profit-On-Cost		1.50x						
Discount Rate		9.6%						
Present Value of Equity Cashflows	20		548,520	512,923	99,203	(106,934)	10,990,686	
Valuation/Appraisal	21	**12,044,399**						
v		7,333,159	(1,065,721)		−14.5%			

Figure 7.2 Example projected levered annual cashflows and DCF valuation.

Source: Authors' calculations.

		Unlevered Asset / Portfolio Beta					
		1.00	**1.10**	**1.20**	**1.30**	**1.40**	**1.50**
	0%	11.816	11.519	11.231	10.953	10.683	10.422
Effective Tax Rate	5%	11.891	11.599	11.317	11.043	10.777	10.519
	10%	11.967	11.681	11.403	11.133	10.871	10.617
	15%	12.044	11.763	11.490	11.224	10.967	10.717
	20%	12.122	11.846	11.577	11.317	11.063	10.817
	25%	12.200	11.929	11.666	11.410	11.161	10.919

Figure 7.3 Valuation sensitivity analysis (millions) illustration.

Source: Authors' calculations.

that applies. Further sensitivity can be conducted on the assumed private vehicle's capital structure and vehicle level considerations such as cost.

7.1.3   *Valuing PREFs*

7.1.3.1   *Key valuation methodologies for closed-ended and open-ended PREFs*

We now extend the valuation methodologies detailed thus far to PREFs. As outlined in Chapter 6, PREFs can be structured in closed-ended and open-ended formats which will have a significant bearing upon the valuation approach undertaken. Closed-ended funds have a defined lifecycle and so should be valued using a DCF approach based on their cash flow projections. This will be illustrated further below. Open-ended funds are "unitised" evergreen vehicles and so can be valued through a DCF-based approach or using some of the methodologies employed to value publicly traded companies (see Chapter 8), for example, the NAV method, depending on an investor's investment horizon or preference.

7.1.3.2   *PREF NAV*

In a similar vein to the EPRA NAV adjustments (referenced in Section 8.3) to recognised accounting standards that better reflect real estate industry considerations, PREFs can also utilise a similar approach. INREV, a European PREF industry body, has provided a NAV basis that reflects several recommended modifications to International Financial Reporting Standards (IFRS). See www.inrev.org for more details and guidance on this. There are several INREV NAV adjustments, with the typical major ones being:

• The reclassification of shareholder loans and other debt or hybrid (e.g., convertible bonds) instruments representing long-term investor interests as equity.
• The fair valuation of real estate assets held including any finance leases or minority indirect investments. This should reference independent appraisals where relevant and available ensuring that the NAV reflects market values; and
• The capitalising and amortisation of one-off costs, including fund setup, direct transaction and loan arrangement costs and fees, over a five-year period. This better represents the "duration of economic benefits" for PREFs, and these can be significant items added to balance sheet assets, for example, direct transaction costs are often 4–6%+ of purchase price and when using leverage can easily represent net costs of 10%+ of equity invested. This is especially impactful in the case of smaller or early-stag lifecycle PREFs,

where these costs represent a greater proportion of equity versus large PREFs transacting small volumes relative to their NAVs.

7.1.3.3 DCF-based valuations

Having made cashflow projections for PREFs, which can be either bottom-up, led by aggregating up individual transaction cashflows, or, taking a more macro approach, using historical cashflow profiles combined with market level forecasts (e.g., Buchner 2017) for a private equity funds example, a discount rate then needs to be applied to conduct a DCF valuation. Referencing the range of estimated PREF betas collated from various studies in Figure 6.2, using the PREF cashflow example in Figure 6.5 and extending the assumptions used in the various valuation exercises in this text, Figure 7.4 provides an illustration of a PREF valuation exercise.

Because we are assuming that this illustrative PREF is newly launched and has yet to draw capital, the discount rate or required return is used here to estimate the net present value (NPV) of the opportunity. A positive NPV illustrates that an investment will be profitable when discounting projected net (positive minus negative) cashflows using a specific discount rate. The impact of including a catch-up mechanism (illustrated in Chapter 6) is also provided and highlights that the additional performance dilution created can have a meaningful bearing on investor decision-making using such a methodology.

To highlight a PREF DCF valuation, Figure 7.5 shows the current value of the net cashflows at the end of Year 4 of the PREF's lifecycle. At this stage, the closing NAV is 120.8. Extending the same approach of Figure 7.1, the valuation is shown for a range of discount rate assumptions. At this stage, a PREF would be well through its lifecycle and likely have completed its investment activity. Thus, an investor may perceive that as reflecting less of an execution risk at that stage and thus assign a commensurately lower required return when compared to expectations at the start of the PREF.

Year	IRR	\multicolumn{8}{c}{Net PREF Cashflows}
		1	2	3	4	5	6	7	8
No Catch-Up	11.6%	-30.8	-31.5	-30.2	-7.4	30.5	69.6	51.0	5.1
With 50:50 Catch-Up	10.7%	-30.8	-31.5	-30.2	-7.4	30.5	69.6	47.0	3.2

Valuation Assumptions

Risk Free Rate	4.0%
Risk Premium	3.0%

		\multicolumn{2}{c}{NPV}
Fund Beta	Discount Rate	No CU	Incl CU
1.00	7.0%	16.3	12.7
1.25	7.8%	13.3	9.8
1.50	8.5%	10.4	7.1
1.75	9.3%	7.6	4.5
2.00	10.0%	5.1	2.1
2.25	10.8%	2.7	-0.2
2.50	11.5%	0.4	-2.3
2.75	12.3%	-1.7	-4.3
3.00	13.0%	-3.7	-6.2

Figure 7.4 Illustrative PREF cashflow and investment worth assessment.

Source: Authors' calculations.

Fund Year	IRR	Net PREF Cashflows							
		1	2	3	4	5	6	7	8
Net PREF Cashflow	11.6%	-30.8	-31.5	-30.2	-7.4	30.5	69.6	51.0	5.1
Closing NAV			30.0	64.2	101.9	122.1	107.4	52.2	5.6	0.0

Valuation Assumptions

Risk Free Rate	4.0%
Risk Premium	3.0%

Present Value of PREF Cashflows At End Year 4

Fund Beta	Discount Rate	1	2	3	4	Valuaton	Vs PREF NAV
1.00	7.0%	28.5	60.8	41.7	3.9	134.9	1.10x
1.25	7.8%	28.3	59.9	40.8	3.8	132.9	1.09x
1.50	8.5%	28.2	59.1	39.9	3.7	130.9	1.07x
1.75	9.3%	28.0	58.3	39.1	3.6	129.0	1.06x
2.00	10.0%	27.8	57.5	38.3	3.5	127.1	1.04x
2.25	10.8%	27.6	56.7	37.6	3.4	125.3	1.03x
2.50	11.5%	27.4	56.0	36.8	3.3	123.5	1.01x
2.75	12.3%	27.2	55.2	36.1	3.2	121.8	1.00x
3.00	13.0%	27.0	54.5	35.4	3.2	120.0	0.98x

Figure 7.5 Illustrative PREF valuation at the end of year four.

Source: Authors' calculations.

7.1.4 Secondary markets and structures

A secondary market has developed for the interests in PREFs and is intermediated by a range of groups, including specialist brokers, investment banks and other financial institutions. The expansion of the PREF market has helped to create the requisite depth of product for a real estate secondaries market.

For vendors, the key rationales for using the secondary market are as follows:

• Short-term liquidity requirements: This could be due to PREFs being unable to provide liquidity and in the case of open-ended funds, it would be due to delayed redemptions owing to large queues or suspensions at times of market disruption.
• Portfolio management: This could be due to a need to reduce overall real estate exposure in a multi-asset context or allocations within a real estate programme.
• Regulatory changes: By way of example, both the Basel III and Solvency II regulations made it onerous in terms of capital reserve requirements for banks and insurers, respectively, to hold illiquid equity investments such as PREFs and this triggered meaningful sales activity upon their implementation.

For investors, there are several benefits, including:

• J-curve reduction/offset: secondaries can negate the need to pay management fees on undrawn capital and remove/reduce exposure to upfront transaction costs in the underlying real estate investments.
• Discounted entry pricing: frequently possible, which augments performance.

- Diversification: provide immediate exposure to an existing pool of assets and for a portfolio transaction, allocations across managers and fund vintage years.
- Reduced execution risk given secondaries entail an investment in funds that are progressed in their lifecycles, investors have visibility on the underlying investments which are specified and have a reduced hold period/duration profile.

Secondaries transactions can also be structured in several ways, including:

- "Vanilla": simple transfer of partial fund interests between two parties. This could be for individual interests or portfolios.
- Recapitalisations: investments in PREFs at or near the end of their lifecycles, providing exiting LPs with proceeds and GPs with additional capital and time to continue managing the underlying investments.
- Deferred payments: a form of leverage for a secondary investor whereby only a partial amount of the consideration is paid upfront with the remainder deferred over future periods. Any interim distributions made can be used to offset any remaining deferred amounts that are payable.
- Directs: instead of fund interests, other indirect investment structures could be transacted, including joint ventures or interests in direct assets.

To illustrate the mechanics of a secondary transaction, a cashflow profile for a "vanilla" secondary in a PREF following the above valuation examples. Here the investor acquires a 10% interest in the PREF at a 15% discount to the prevailing NAV at the end of year four (Figure 7.6).

As can be seen, the combination of discounted entry pricing and the abridged timeframe of a four-year-old positive performance is positively amplified to an expected IRR of 20.8%. However, due to the shortened timeframe of four years when compared to the PREF's eight-year projected life, the forecast equity multiple is lower for the secondary transaction, which is a common feature of these investment types. Figure 7.7 highlights the performance sensitivity from varying levels of discounted entry pricing. As can be seen, an approximate 18% discount to NAV entry price would be required to equate with the projected equity multiples.

Net Fund Cashflows								
Primary Year	1	2	3	4	5	6	7	8
No Catch-Up	-30.8	-31.5	-30.2	-7.4	30.5	69.6	51.0	5.1
Closing NAV	30.0	64.2	101.9	122.1	107.4	52.2	5.6	0.0
IRR	11.6%							
Equity Multiple	1.6x							

Secondary Cashflows					
Secondary Year		0	1	2	3	4
Secondary Interest	10.0%	-12.2	3.1	7.0	5.1	0.5
Secondary Pricing	15.0%	-10.4	3.1	7.0	5.1	0.5
IRR	21.1%					
Equity Multiple	1.5x					

Figure 7.6 Illustrative secondary transaction cashflows.

Source: Authors' calculations.

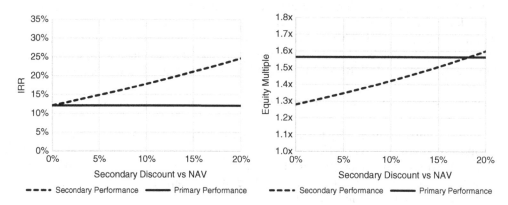

Figure 7.7 Illustrative secondary transaction projected performance sensitivity versus entry pricing.

Source: Authors' calculations.

Finally, in applying the same DCF valuation methodology used in Figure 7.5, Figure 7.8 highlights how a secondary purchaser may choose to appraise their holding. Simple approaches can involve treating the secondary price as market value or the investor can reference the PREF's current NAV, thereby recording a capital gain arising from the

		Net Secondary Cashflows				
Secondary Year		0	1	2	3	4
Secondary Interest	10.0%	-12.2	3.1	7.0	5.1	0.5
Secondary Price	15.0%	-10.4	3.1	7.0	5.1	0.5
IRR	21.1%					
Equity Multiple	1.5x					

Valuation Assumptions

Risk Free Rate	4.0%
Risk Premium	3.0%

		Present Value of Secondary Cashflows						
Fund Beta	**Discount Rate**	1	2	3	4	**Valuaton**	**Vs Secondary Price**	**Vs Fund NAV**
1.00	7.0%	2.9	6.1	4.2	0.4	13.5	1.30x	1.10x
1.25	7.8%	2.8	6.0	4.1	0.4	13.3	1.28x	1.09x
1.50	8.5%	2.8	5.9	4.0	0.4	13.1	1.26x	1.07x
1.75	9.3%	2.8	5.8	3.9	0.4	12.9	1.24x	1.06x
2.00	10.0%	2.8	5.8	3.8	0.4	12.7	1.23x	1.04x
2.25	10.8%	2.8	5.7	3.8	0.3	12.5	1.21x	1.03x
2.50	11.5%	2.7	5.6	3.7	0.3	12.4	1.19x	1.01x
2.75	12.3%	2.7	5.5	3.6	0.3	12.2	1.17x	1.00x
3.00	13.0%	2.7	5.5	3.5	0.3	12.0	1.16x	0.98x

Figure 7.8 Illustrative secondary transaction DCF valuation.

Source: Authors' calculations.

secondary price mark-up to the NAV. Using a DCF approach, the discount rate selected will determine the valuation relative to either the NAV or the secondary price.

Quiz questions

1 What considerations should investors factor into discount rates when valuing private real estate vehicles?
2 What are the calculated vehicle risk premia for the following leverage and tax assumptions assuming a market risk premium of 5%?

   a Unlevered beta = 1.1x, tax rate = 20%, LTV = 50%
   b Unlevered beta = 1.25x, tax rate = 40%, LTV = 20%
   c Unlevered beta = 1.5x, tax rate = 10%, LTV = 75%

3 For a PREF with an existing portfolio that has just been expanded through a new acquisition using debt finance, calculate this PREF's NAV at the current time and in year three with and without INREV NAV adjustments using this simple balance sheet format.
   Property Assets +
   Debt −
   Other Net Assets/Liabilities +/−
   INREV Adjustments +/−
   Net Asset Value

Assumptions to use:

- Current property assets = 100 (fair market value)
- New acquisitions = 200 (fair market value)
- Capital growth = 3% p.a.
- Acquisition costs = 10%
- Debt employed = 100 (no change to this)
- Loan arrangement costs = 2%
- Other net assets/liabilities = +5 (no change to this)

Research/Dissertation topics

- Estimating discount rates and other valuation considerations for private real estate platforms operating in non-traditional commercial real estate sectors, for example, self-storage. This would be well suited to case studies.
- How do the valuations of PREFs evolve through their lifecycle as their risk profile shifts?

References

Buchner, A. (2017). Risk management for private equity funds. *Journal of Risk*, 16 (6): 1–32.
Farrelly, K. & Stevenson, S. (2019). The risk and return of private equity real estate funds. *Global Finance Journal*, 42, 1–13.

8 Private commercial real estate debt structures

- Introduction
- Commercial real estate mortgages
- Junior or mezzanine loans
- Current regulatory environment
- Quiz

Objectives

At the end of this chapter, you will understand:

- The basic features of private commercial real estate mortgages.
- The economics and cashflows received by lenders.
- How lenders underwrite and mitigate risks.
- The impact of current regulations.

8.1 Introduction

Commercial real estate debt is used by investors seeking to acquire and own privately held properties. Investors do this to adjust the risk-return profile of the opportunity and/or because they have insufficient equity capital. Debt can be used to complete asset acquisitions and finance capital or development expenditures. Secured real estate debt is legally collateralised by a specific asset or portfolio, whereas unsecured debt is provided to real estate companies, funds, or other entities at the vehicle level. In both cases, debt is recognised as a long-term liability on balance sheets and lenders receive a stream of cashflows to compensate them for the risk that they take.

Other reasons for the use of private debt capital in commercial real estate investments include managing the overall cost-of-capital and garnering tax benefits through the tax deductibility of interest payments (please see Chapter 7). Several finance theories also provide rationales, including:

- Modigliani-Miller: no justification
- Trade-off theory: optimal leverage level which maximises return in the presence of tax incidence.
- Pecking order: easier to raise debt capital than equity capital.
- Market timing: raise debt when debt is cheap and equity returns are attractive.
- Incentive theory: management motivated to grow business and enhance remuneration.

DOI: 10.1201/9781003298564-10

- Industry effects: herding towards industry average leverage levels, for example, private real estate fund risk styles such as Core, which has defined leverage levels, could drive this.

8.2 Commercial real estate mortgages

8.2.1 Basic features of a private commercial real estate mortgage

Commercial real estate mortgages are the most common form of debt finance used by commercial real estate investors in the asset class. These are typically secured in income-producing properties, providing the income to service the interest payments, but are also provided to fund capital-intensive projects such as developments. In both instances, there is a security interest in both the land and property.

Debt providers, known as "mortgagees" or "lenders", are commonly financial institutions such as banks and insurance companies, with an increasing range of 'alternative' lenders appearing, including dedicated private funds and listed entities (e.g., mortgage REITs). In the event of default (to be discussed in more detail), lenders have the right to enforce a foreclosure process through the judicial system to ultimately take ownership of the property directly. It is also customary that borrowers will pay all due diligence expenses incurred by the lender.

Mortgages can be made on a recourse or non-recourse basis. In the event of borrowers defaulting on their loan obligations, non-recourse mortgages only enable the lender to recover proceeds from the sale of the reference collateral. In the case of recourse loans, where borrowers are all required to provide additional collateral pledges, the lender will have the right to additional assets of the borrower where the underlying asset sales proceeds are insufficient to cover the mortgage balance. Further credit enhancement for lenders can include borrowers providing personal guarantees, making them a legal requirement to pay the required proceeds in the event of default or if interest payments go into arrears.

In return for providing mortgage finance, lenders receive a combination of fees and interest payments as cashflows. Fees can include upfront arrangement and exit fees, defined as a percentage of the loan balance provided. Periodic interest payments are calculated by a combination of a reference interest rate, for example, a swap rate or government bond yield, and an additional spread referred to as interest margin, which compensates the lender for the risk entailed on the loan.

To illustrate the level of commercial real estate interest margins and how they can evolve, Figure 8.1 shows the evolution of UK margins across several property types. As can be seen, there is a cyclical dimension to these, with a significant increase in the years post the GFC period, when the pricing of real estate rebased significantly. In more recent years, margins have shown a marked divergence across sectors, with prime offices and industrial margins property types being priced approximately 140 basis points lower than secondary offices and hotels.

Commercial real estate mortgages have fixed terms and can be structured on a floating or fixed interest rate basis. Borrowers may also wish or be required by lenders to utilise interest rate hedging derivative instruments such as swaps or options to either fully fix or lock in a range of interest costs. They can also be amortising or non-amortising. Mortgages that are not amortising are often referred to as bullet or interest-only loans, where the full principal balance of the loan is required to be repaid upon maturity. Amortising loans instead require the borrower to pay down the loan balance over its term, which is an additional payment to the interest cost (interest rates plus a margin)

Figure 8.1 Historical average UK senior commercial real estate loan margins by sector.

Source: Bayes Business School.

obligations. Fully amortising loans are ones which fully repay the loan through the term while partially amortising loans are those which only result in a partial repayment. The level of amortisation is dictated by the lender's risk-return requirements.

8.2.2   *Senior commercial mortgage cashflows*

To illustrate the economic returns to a senior commercial real estate mortgage lender, namely a provider of a single loan, Figure 8.2 shows summary cashflows. Here a commercial property costing 10 million is purchased by an investor using a 5 million loan million with an annual cost of 4.5% (comprised of a reference interest rate plus a margin). This loan has a term of 20 years and is fully amortising over this timeframe, with the interest cost fixed throughout the term. In each month, the amortisation payment reduces the balance of the loan until expiry, when it is fully repaid. The 'mortgage loan constant' concept represents the monthly payment that would be required to repay a loan given its stated interest rate and term. This payment is 31,632 per month and includes both interest and amortisation components. See the accompanying Excel files for all the calculations.

Over the term of the loan, the interest payment scales down such that by the latter stages, it represents a fraction of the total cost. This can be seen in Figure 8.2, where we show just the first ten and the final ten months of cashflows. The reason for this is that the level of interest to be paid each month is based on a fixed percentage of a declining balance and, conversely, the amortisation component of the required monthly payment grows over time. This is highlighted graphically in Figure 8.3 where, on the left-hand side, we show the monetary split of the fixed monthly payment and on the right-hand side, the pro rata contributions. The interest payments declined steadily from an approximate 60% share to 30% after 12 years. In this example, the loan had a 20-year term, but it is

Start Date	31-Mar-22
Purcahse Price	10,000,000
Loan-to-Value	50%
Loan Amount	5,000,000
Annual Interest Cost	4.50%
Loan Term	20.0
Monthly Loan Payment	31,632

Month	Date	Loan Balance	Interest	Amortization	Total Payment	Closing Loan Balance	Cumulative Interest	Lender Cashflow	Impact of monthly compounding
0	31-Mar-22	5,000,000				5,000,000		-5,000,000	4.594%
1	30-Apr-22	5,000,000	18,750	12,882	31,632	4,987,118	18,750	31,632	
2	31-May-22	4,987,118	18,702	12,931	31,632	4,974,187	37,452	31,632	
3	30-Jun-22	4,974,187	18,653	12,979	31,632	4,961,207	56,105	31,632	
4	31-Jul-22	4,961,207	18,605	13,028	31,632	4,948,180	74,709	31,632	
5	31-Aug-22	4,948,180	18,556	13,077	31,632	4,935,103	93,265	31,632	
6	30-Sep-22	4,935,103	18,507	13,126	31,632	4,921,977	111,772	31,632	
7	31-Oct-22	4,921,977	18,457	13,175	31,632	4,908,802	130,229	31,632	
8	30-Nov-22	4,908,802	18,408	13,224	31,632	4,895,577	148,637	31,632	
9	31-Dec-22	4,895,577	18,358	13,274	31,632	4,882,303	166,996	31,632	
10	31-Jan-23	4,882,303	18,309	13,324	31,632	4,868,980	185,304	31,632	
231	30-Jun-41	309,897	1,162	30,470	31,632	279,427	2,586,527	31,632	
232	31-Jul-41	279,427	1,048	30,585	31,632	248,842	2,587,575	31,632	
233	31-Aug-41	248,842	933	30,699	31,632	218,143	2,588,508	31,632	
234	30-Sep-41	218,143	818	30,814	31,632	187,328	2,589,326	31,632	
235	31-Oct-41	187,328	702	30,930	31,632	156,398	2,590,029	31,632	
236	30-Nov-41	156,398	586	31,046	31,632	125,352	2,590,615	31,632	
237	31-Dec-41	125,352	470	31,162	31,632	94,190	2,591,085	31,632	
238	31-Jan-42	94,190	353	31,279	31,632	62,911	2,591,438	31,632	
239	28-Feb-42	62,911	236	31,397	31,632	31,514	2,591,674	31,632	
240	31-Mar-42	31,514	118	31,514	31,632	0	2,591,793	31,632	
							Lender IRR	4.59%	

Figure 8.2 Illustrative senior commercial real estate mortgage cashflows.

Source: Authors' calculations.

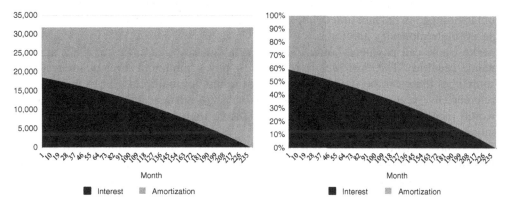

Figure 8.3 Illustrative interest and amortization contributions.

Source: Authors' calculations.

commonplace for commercial real estate mortgages to have shorter terms accompanied by longer amortisation periods. At the expiry of a loan, the borrower would simply be required to repay the reduced outstanding balance.

Finally, the lenders expected cashflow and calculated performance are shown. As can be seen, the expected IRR or yield-to-maturity, as this is often termed, is higher than the annual interest cost. The reason for this is that the 4.5% interest cost is a contracted annual rate and instead, the rate of return is based on a compound rate. The following formula can be used to reconcile the difference between the two:

Compound Interest = (1 + Annual Rate/Compound Period) $\wedge$ *(Compound Period) − 1*

In this example:

4.59% = (1 + 4.50%/12) $\wedge$ *12 − 1*

Now the compound rate fully aligns with the lender's calculated IRR. If a 1% arrangement fee was also incorporated into the lender's economics, here a 50,000 upfront payment, then the estimated IRR would increase to 4.7%, a marginal increase owing to the long duration (20 years) of the loan's term.

8.2.3   Lender considerations

8.2.3.1   Prepayment protections

Commercial real estate mortgages commonly allow the borrower to repay the loan in advance of its maturity date. This prepayment risk that lenders are exposed to can be negated through the use of penalties, which ensure a minimum level of financial return for the time, resources and capital committed by the lender. These prepayment protections can include:

- Lockout periods: these are pre-specified periods in which borrowers are prohibited from repaying the loan either partially or in full.
- Monetary prepayment penalties: these are calculated by reference to the prevailing outstanding loan balance and by multiplying it by a specified rate, or it can be stated as an absolute monetary amount.
- Make-whole call payments: often referred to as interest guarantees or yield mainte-nance penalties. The borrower is required to pay a minimum level of interest and sometimes these can be stated in the form of minimum multiples. For example, for a reference five-year mortgage, a borrower may wish to repay at the end of year two but would then be required to pay a charge to ensure that they paid three years of total interest payments.
- Defeasance: potential method available to borrowers to avoid paying fees to lenders. Instead, the borrower is able to swap the property asset or portfolio for another cash-flowing asset, typically fixed-income securities, as collateral for the loan to support the interest payments.

To show the impact on lender returns, Figure 8.4 assumes a commercial real estate mortgage with a five-year term and a 20-year amortisation schedule. The borrower is

Start Date	31-Mar-22		Make Whole	3	Years Interest
Purcahse Price	10,000,000		Min. Interest	643,231	
Loan-to-Value	50%		Prepayment Year	2	
Loan Amount	5,000,000		Interest Paid	436,293	
Annual Ineterst Rate	4.5%		Prepayment Fee Due	206,938	
Amortization Term	20.0				
Monthly Loan Payment	31,632				

Month	Date	Loan Balance	Interest	Amortization	Total Payment	Closing Loan Balance	Cumulative Interest	Full Term Lender Cashflow	Two-Year Hold Lender Cashflow
0	31-Mar-22	5,000,000				5,000,000		-5,000,000	-5,000,000
1	30-Apr-22	5,000,000	18,750	12,882	31,632	4,987,118	18,750	31,632	31,632
2	31-May-22	4,987,118	18,702	12,931	31,632	4,974,187	37,452	31,632	31,632
3	30-Jun-22	4,974,187	18,653	12,979	31,632	4,961,207	56,105	31,632	31,632
4	31-Jul-22	4,961,207	18,605	13,028	31,632	4,948,180	74,709	31,632	31,632
5	31-Aug-22	4,948,180	18,556	13,077	31,632	4,935,103	93,265	31,632	31,632
20	30-Nov-23	4,746,794	17,800	13,832	31,632	4,732,962	365,611	31,632	31,632
21	31-Dec-23	4,732,962	17,749	13,884	31,632	4,719,078	383,360	31,632	31,632
22	31-Jan-24	4,719,078	17,697	13,936	31,632	4,705,142	401,057	31,632	31,632
23	29-Feb-24	4,705,142	17,644	13,988	31,632	4,691,154	418,701	31,632	31,632
24	31-Mar-24	4,691,154	17,592	14,041	31,632	4,677,113	436,293	31,632	4,915,684
25	30-Apr-24	4,677,113	17,539	14,093	31,632	4,663,020	453,832	31,632	
55	31-Oct-26	4,230,499	15,864	15,768	31,632	4,214,731	954,517	31,632	
56	30-Nov-26	4,214,731	15,805	15,827	31,632	4,198,904	970,322	31,632	
57	31-Dec-26	4,198,904	15,746	15,887	31,632	4,183,018	986,068	31,632	
58	31-Jan-27	4,183,018	15,686	15,946	31,632	4,167,071	1,001,755	31,632	
59	28-Feb-27	4,167,071	15,627	16,006	31,632	4,151,065	1,017,381	31,632	
60	31-Mar-27	4,151,065	15,566	16,066	31,632	4,135,000	1,032,948	4,166,632	
							Lender IRR	4.59%	6.69%

Figure 8.4 Illustrative senior commercial real estate mortgage with a make-whole penalty payment.
Source: Authors' calculations.

repaying this loan at the end of year two and is required to make-whole three years of interest receipts, which would have totalled 643,231. At that time the borrower had only paid 436,293 and thus the borrower was required to make a penalty payment representing the difference between these two amounts in addition to the remaining balance of the loan. This is highly accretive to the lender's rate of return due to the shortened hold periods when they received an additional year of interest.

Borrowers can have many reasons for seeking to fully repay loans ahead of the maturity date, for example, if there have been significant market interest rate movements making new mortgages materially cheaper or due to business plan completion and an earlier than envisaged sale. It is commonplace for lenders to seek protections. Whilst historical data is challenging to source, buoyant market conditions and high debt availability, that is, periods of competitive tension amongst lenders, can lead to these protections being more borrower-friendly and vice versa.

8.2.3.2   *Key metrics and covenants*

Lenders reference key metrics when assessing the risks entailed in the loans that they are underwriting. Ultimately, a lender needs to be comfortable that the borrower is able to meet their interest payments and that the value of the assets will remain higher than that of the mortgage balance.

The first of these is the LTV, which is simply the ratio of the loan balance to the property or portfolio value. To inform this, an independent appraisal move typically is required, which may differ from a purchase price. The lower the LTV, the less risky the commercial mortgage will be and, as a consequence, the interest margins will correlate with this. Over the term of the loan, borrowers will be required to test this through periodic independent appraisals, which typically occur at least annually.

Another key ratio is the debt service coverage ratio, or interest coverage ratio as it is often known. This is the annualised or periodic NOI divided by the equivalent interest cost. A ratio value of one means that the NOI is fully servicing the interest costs, although in practice, ratios comfortably above one is typically sought by lenders, with the assessed security of income being the key driver of this.

Debt yields are calculated using the commercial real estate yield formulas outlined in Chapter 4, with the debt balance substituted into the formulas instead of the property value. This correlates closely to the debt service coverage ratio and LTV ratio but provides another reference point for lenders.

These metrics enable lenders to judge how much they should lend and also inform the "rating of loans" from a risk perspective. Credit ratings are regularly provided by third parties or are estimated by lenders, and these align with those shown in Chapter 11 for public real estate bonds.

Commercial real estate loans normally include several covenants using these key metrics as limits/thresholds, which are monitored continuously. Should one or a number of these covenants not be met before the expiry of the loan, the legal documentation will empower the lender to act. These can include several steps, with the ultimate resolution being a legal enforcement of the property or portfolio and the lender taking ownership through a judicial process. Intermediate steps can include the borrower providing additional capital to pay down the loan in the case of an LTV breach or a "cash-sweep" being triggered, whereby the excess cashflow generated by the property, having met its interest cost obligation, accrues to the lender in repaying the loan balance.

8.3   Junior or mezzanine loans

The discussion in this chapter has focused on single commercial real estate loans, often referred to as senior loans, but the capital structure for a commercial real estate investment can also include additional loans that are subordinate to the senior lender. Figure 8.5 gives a simple overview of two potential capital structures for a reference property investment. The first of these is capitalised by a single senior loan and equity whereas the second capital structure shows a junior or mezzanine loan, which would require the investor to provide less equity capital to fund the transaction.

The senior lender continues to have a first charge or legal priority on the underlying collateral, but the junior lender will have a second charge. Cashflow and repayment of proceeds are firstly prioritised to the senior lender and secondly to the mezzanine lender, and only then does the equity investor receive all residual cashflows. The use of mezzanine

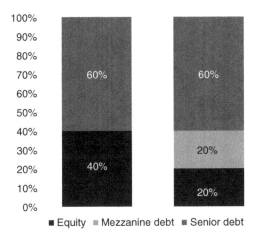

Figure 8.5 Two commercial real estate capital structures.

Source: The authors.

loans entails higher leverage for the equity investor, which in turn increases the risk-return profile of the investment in question.

Mezzanine loans can be structured on an interest or coupon-only basis, or mezzanine lenders can also participate in the performance upside through profit participation. Such lenders also typically have more stringent covenants than the senior lenders, which enables them in turn to have more time to respond in the event the underlying real estate collateral underperforms. These loans are now commonly provided by alternative lenders, such as dedicated private funds, due to their higher risk and the regulatory burdens now placed on both banks and insurers, making them inefficient investments for these institutions.

8.4 Current regulatory environment

Since the GFC, when there was widespread distress in the financial system globally, there have been increasing regulations bearing down to ensure that they have the ability to absorb losses at times of economic and market distress. These have ultimately curtailed the risk-taking activities of these groups, which has had a significant impact on commercial real estate lending. Two examples of these are as follows:

• **Solvency II:** This applies to EU and UK insurers and is a regulatory programme that was introduced in January 2016, governing their investment activities by placing solvency capital requirements (SCRs) on them. The SCR dictates the funds that they are required to hold in order to withstand the losses anticipated in a "tail risk" 99.5% scenario. Varying asset classes are assigned capital charges, and senior real estate loans enjoy a highly favourable treatment. This is ultimately due to the recourse to underlying real estate assets and the covenant protections that lenders can have, which has manifested as low historical defaults and loss rates, especially for investment grade rated loans. This treatment varies according to their inherent risk, as measured by LTVs and the nature of the underlying collateral, but lower-risk investment grade commercial real estate mortgages have a particularly low SCR compared to alternatives and have a healthy duration profile for liability matching.

Figure 8.6 Historical UK senior commercial real estate average maximum LTVs by sector.
Source: Bayes Business School.

- **Basel III/IV:** In a similar vein to Solvency II, these regulations which have evolved, and are set by the Bank of International Settlements, place risk-weightings on the investments that global banks hold on their balance sheets. For commercial real estate lending, the LTV is a key driver of the capital requirements for loans, whether the loans are for land acquisitions, construction, or development activities. Basel IV will provide a more standardised way for banks to have to calculate risk versus being more reliant upon internal models, essentially levelling the playing field across the industry.

The impact of these regulations has been significant and has had the desired effect in terms of curbing risk-taking. This is evidenced by the trends highlighted in Figure 8.6, which shows that the average LTV for UK senior commercial real estate mortgages has reduced meaningfully since the GFC period. In recent years the level has moderated to a 50–60% range across property types compared to 70–80% in the 1999 to 2006 period. This is a level beyond which the regulatory capital requirements for either banks or insurers become relatively capital expensive and, thus, the return on regulatory capital becomes relatively unattractive.

Quiz questions

1 What are the key economic drivers of private commercial real estate loans?
2 Calculate the monthly and quarterly interest payments for the following commercial real estate loans.

   a 50 million size, 10-year term and 5.0% interest rate
   b 10 million size, 10-year term and 6.0% interest rate

   c 50 million size, 10-year term and 5.0% interest rate
   d 50 million size, 10-year term and 4.0% interest rate

3 For all of the above commercial real estate loans in Question 2, calculate the compound interest rates for the monthly and quarterly payment periods.
4 Using the model for Figure 14.4 provided in the accompanying Excel file, calculate the lender's IRR or yield-to-maturity for the following two prepayment penalty scenarios:

   a The make-whole penalty increases to 3.5 years of interest payments.
   b The lender instead is guaranteed a 1.2x profit on their capital.

Research/Dissertation topics

- The inflation hedging attributes of private commercial real estate loans.
- A risk-return assessment of commercial real estate mezzanine loans
- How can lenders incorporate ESG considerations into their underwriting?

Public markets

9 Public market structure

Objectives

At the end of this chapter, you will understand:

- The key participants in the market, their roles, remuneration sources and motivations.
- The difference between the buy side and the sell side, as well as the importance of the Chinese Wall.
- The role, use, and importance of formal (index) and informal (peer group) benchmarks.
- How consensus forecasts and market expectations are formed and their role in determining share prices.
- Key trade associations and data providers in the sector.

Key concepts

- Interaction of market participants.
- Different motivations of the buy and sell side.
- The role of benchmarks.
- The expectations and forecasts implicit in share prices.

9.1 Who are the key players?

In this first section, we supply an outline of the key participants in the listed sector, grouped according to function, and examine their roles and motivations. The key point to understand is how the different roles affect decision-making and how, collectively, their actions affect real estate securities prices. Figure 9.1 provides a simple framework for understanding the separate roles. It is divided into five elements, four in each quadrant, with the fifth being an individual company at the centre of the interaction.

As can be seen, there are two segments to be borne in mind: Buy Side (left-hand top quadrant) vs. Sell Side (right-hand top quadrant), and Above (top half) vs. Below (bottom half) the Chinese Wall. It is important to clarify what is meant by these terms before moving on to explore the individual roles.

The buy side refers to institutions which manage assets (be they real estate, equities, funds, bonds, or commodities) either on their own behalf or on behalf of others. They are therefore acting as principals. Typical buy-side institutions are pension funds, sovereign wealth funds, private equity funds and asset managers.

DOI: 10.1201/9781003298564-12

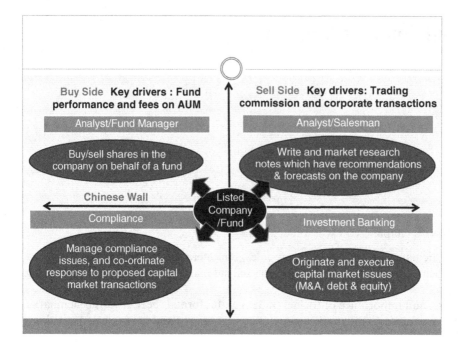

Figure 9.1 Listed market participants by role.

Source: Consilia Capital www.consiliacapital.com.

The sell side refers to institutions who provide services (research, execution, advisory) to the buy side in return for trading commission, advisory or transaction fees. Typical sell-side institutions are stockbrokers, investment banks and real estate agents.

Although firms are typically thought of as being either buy-side or sell-side, confusingly, one institution can have both buy-side and sell-side activities within the same organisational umbrella (albeit separated for compliance purposes). Examples are Goldman Sachs, JP Morgan, and Morgan Stanley. This chart can therefore represent both the division between different firms and the division between different departments of the same firm.

A Chinese Wall is defined by the Corporate Finance Institute[1] as "a virtual information barrier erected between those who have material, non-public information, and those who don't, to prevent conflicts of interest". For our purposes, the key point to remember is that it is only legally possible to trade with the benefit of publicly available information. This ensures that no participants in the market have an information advantage over others. Once one is aware of this, it becomes obvious why companies now have such comprehensive presentations/investor updates at the time of the release of their figures and throughout the year. Their aim is to ensure, where possible, that all key information is made publicly available; therefore, the share price should be an accurate reflection of the company's prospects.

9.2   Buy-side and sell-side motivations

Looking at the individual quadrants, it is possible to provide a summary of their motivation, actions, and impact on the listed sector in order to compare and contrast,

	Sell-side Analyst	Buy-side Analyst	Corporate Finance	Compliance
Motivation	Trading Commission, Research Fees (post-MiFID II), and (indirectly) related corporate finance fees from individual companies researched.	Fund outperformance vs. benchmark, Asset Management fees (linked to size of fund and performance)	Fees on transactions	Avoid conflicts of interest or insider dealing
Actions	Research notes with recommendations and forecasts on individual companies, along with reports on the industry	Buy/sell securities to adjust weightings, paying for brokers' research either themselves or via trading commissions.	Issuance of securities/bonds plus M&A	Co-ordinating lists of securities which cannot be traded.
Impact	Collectively (i.e., aggregating all sell-side analysts) these form the "consensus" view	Collectively these investors will underwrite future equity issues.	Liaising with existing and new shareholders	Act as a conduit for corporate finance

Figure 9.2 Summary of key participants' motives and actions.

Source: The authors.

as well as understand how they interact around the central axis of the listed company. This is shown in Figure 9.2. We will look at their actions in more detail in a case study later: "A Day in the life" – what happens when a company releases its figures.

In addition to these four main groups of participants, there are also the roles specific to one company, both internal and external, such as the corporate lawyers, accountants, PR, and IR teams.

9.3 Benchmarking

Now that the market participants have been divided into different groupings which reflect their motivations and aims, it is important to look at some of the key concepts involved in the trading and performance measurement of listed real estate securities, such as benchmarking and consensus estimates. This will lead us to our second group of participants, the information and service providers. In this section, we focus on the concept of benchmarking.

There are two ways of looking at performance measures, first, in an absolute sense and, second, in a relative sense. Let us take an example.

Case study: LandSec results

The NAV per share (A key investment metric dealt with in detail later in this book) of UK REIT LandSec for the six months to September 2021 rose by 2.7% to 1012p per share. Three questions arise from this:

1 *Is this a good or bad result?*
2 *What would you expect the share price to do on the announcement of these figures?*
3 *Are fund managers likely to increase or decrease weightings because of this added information?*

We cannot answer these questions with just this absolute information. We need to know:

1 *How this performance compares with other similar portfolios (benchmarking) and*
2 *How this performance relates to expectations (consensus estimates dealt with in the next section).*

In terms of benchmarks, we can split the use into two categories, formal (e.g., Index) vs. informal (peer group) and company vs. fund manager use.

We now explain how they are constructed and then look at how they are used in practice.

9.3.1 Formal benchmarks

These can be based either on the market capitalisation of the company or the gross assets of the portfolio. For the listed real estate sector, the best-known series is the FTSE/EPRA/NAREIT series of indices.

9.3.2 Equity benchmarks

Figure 9.3 shows the weightings of the largest real estate companies globally. These are the most "investible" companies at that time (December 2021). For each company, we show the size (free-float market capitalisation) and the weight in the index. What is the importance of "free float" in this context? The index market capitalisation takes the gross market capitalisation and takes off the shareholdings of "insiders", be they directors or strategic long-term shareholders, who will not trade their shares in the market. This therefore represents the number of shares that can be bought or sold by investors.

9.3.3 Property benchmarks

In certain regions, such as Europe, companies use the IFRS for accounting purposes. In others, such as the US, they use local Generally Accepted Accounting Principles (GAAP). This is important for several reasons, one of the main ones being that under IFRS, a company has to revalue its assets and incorporate that valuation into the balance sheet. Real estate investment assets are therefore not depreciated and are held at market value. This means that the company's NAV reflects the market value of assets. As a result, the change in the NAV is one of the key metrics used to analyse and judge the performance of a listed real estate company in Europe. This explains why the discount/premium to a NAV valuation metric is so popular. In contrast, in the US, a company holds its assets at cost less depreciation. The company's NAV therefore stands for its book value, not its market value. As a result, a discount/premium to NAV is a less common, but not unused, investment

Company Name	Free Float Mkt Cap (EUR m) Dec-21	Free Float Index Weight (%) Dec-21
Prologis	108,968.25	5.98
Equinix	66,375.56	3.64
Public Storage	49,560.73	2.72
Simon Property	45,731.04	2.51
Digital Realty Trust	43,562.99	2.39
Vonovia	37,664.98	2.07
Welltower	35,539.99	1.95
AvalonBay Communities	32,775.60	1.8
Alexandria Real Estate	30,177.60	1.65
Equity Residential	29,268.19	1.61
Extra Space storage	26,087.97	1.43
Invitation Homes	23,741.91	1.3
Mid-America Apartment Communities	23,225.07	1.27
Duke Realty Corporation	21,893.32	1.2
Sun Communities	21,101.48	1.16
SEGRO	20,446.30	1.12
Essex Property Trust	20,026.58	1.1
Ventas	17,873.40	0.98
Healthpeak Properties	17,069.99	0.94

Figure 9.3 Largest companies in the EPRA Global Developed Index by free-float market capitalisation and weighting.

Source: p. 52, EPRA Monthly Statistical Bulletin, December 2021.

	Capital Return	Total Return
Retail	100	120
Offices	400	280
Total	200	140

Figure 9.4 British Land property performance relative to an IPD benchmark shown as outperformance for the five year period in bps.

Source: British Land investor presentation, March 2016.

metric. For a company that does revalue its assets, we can look at a benchmark of how it has performed relative to an underlying real estate benchmark, such as MSCI (formerly IPD). The example in Figure 9.4 shows how UK REIT British Land presented its relative real estate performance.

Case Study 2 Relative Performance to a Property Benchmark

9.3.4 Informal benchmarks – peer groups

When comparing the performance of companies, both management teams and fund managers will assemble what is known as a peer group. These are collections of companies

Style	Factor	Yield	LTV	Analyst Opinion	1m Perf	Volatility
1	Cities	1.8%	24.3%	3.0	3.5%	20.8%
2	Diverse	3.7%	38.5%	1.9	1.0%	30.2%
3	Meds	3.3%	34.3%	4.6	2.2%	21.2%
4	Resi	3.3%	41.5%	2.5	2.8%	27.1%
5	Sheds	2.7%	38.2%	3.0	-1.0%	67.2%
6	Shops	4.6%	40.2%	2.4	2.4%	28.2%
7	Infra	1.7%	29.7%	3.7	1.2%	35.9%
8	Workspace	3.6%	34.6%	2.2	1.8%	26.4%
9	Storage	1.8%	24.5%	4.0	6.1%	22.7%
10	Travel	1.6%	49.0%	1.7	4.7%	29.3%
Bayes Benchmark		3.6%	38.4%			31.1%

Figure 9.5 Asset level analysis.

Source: Bayes Business School Monthly Monitor December 2021.

that are typically in the same asset grouping, for example, US industrials, Paris offices, German residential, and Hong Kong retail. Peer groups can also include companies of comparable size as measured by market capitalisation. This peer group will be used on an ongoing basis for relative performance considerations. Most sell-side analysts bracket the companies they follow into these groups. These groups, or clusters as some fund managers call them, can then be compared to each other to identify key trends. Figure 9.5 shows an example of this approach, where we have grouped (in this case, European) stocks with similar underlying assets and share price drivers. Rather than formal terms, we use investor grouping names such as "Meds" for healthcare assets, "Workspace" for offices, etc. In this example, we show the dividend yield, gearing level, analyst view, one-month performance and volatility.

9.3.5   Company vs. fund manager use

Both companies and fund managers will focus on similar metrics for peer group comparison, albeit with a slightly different emphasis due to their roles. These metrics typically include:

Asset performance.
Forecast growth rates in NAV, dividend per share (DPS) and adjusted funds from operations (AFFO)/earnings per share (EPS).
Executive remuneration.
Cost ratios (admin expenses as a % of rents).
Gearing.
Valuation.
Share price performance.
Shareholder structure.

9.4   Consensus views and market expectations

We now turn to the next key concept needed to understand real estate securities price movements, that of consensus views and market expectations. What process is involved in deciding whether a set of results is good or bad? We can look at this as a four-step process, comprising:

1 Individual company guidance
2 Peer group comparables
3 Individual analyst forecasts
4 Consensus analyst forecasts

9.4.1   Individual company guidance

Across all regions, companies will, to a greater or lesser extent, provide some guidance as to their expected financial performance. In the US, companies provide quarterly financial information; in Europe, it is semi-annual reporting, often with 1st Quarter and 3rd Quarter trading updates. These are the regulatory disclosures. In addition, a company can make discretionary disclosures around portfolio performance, letting progress and disposals. In addition, immediately after the end of a financial period and before the publication of the figures, it can provide guidance as to whether the results will be materially above or below expectations.

9.4.2   Time frame

The main periods used for forecasting listed companies (not just real estate companies) are the full financial year period, not interim (semi-annual) or quarterly periods. This irons out any seasonal variations in the figures. Sell-side analyst reports contain forecasts of one, possibly two, years of historical annual information and two possibly three years of forecasts. It is important to understand that this is always a rolling figure, as companies will report after the year-end. Interim or quarterly updates are only a guide to how the full-year figures are likely to be achieved. Even though there is not the same level of seasonality in a real estate company as say a retailer who books, say, 80% of revenues in the winter holiday period, it is important to always deal in full-year forecasts.

9.4.3   Individual analyst forecasts

Each sell-side analyst will produce a detailed financial model for each company they follow. We will deal with the components and mechanics of that model later in the book. For now, it is sufficient to understand the summary data produced and to illustrate the point made above regarding full-year figures. As can be seen, the table shows a one-year historical and a three-year forecast. A company normally releases its figures six to eight weeks after the period ends. Thus, the figures for March are produced in May, possibly June. The period between the financial year's end and the publication of the figures is known as the close period. When the figures are produced, the analysts alter their forecasts based on the historical figures reported (which represent a base level) and the outlook and trading statement comments from the company (Figure 9.6).

Year-end 31 Mar		2021A	2022F	2023F	2024F
NOI (Rents)	EURm	335	325	352	375
Admin Exes.	EURm	25	26	28	29
Net interest	EURm	125	135	142	160
Reported Profit	EURm	185	164	182	186
NAV per share	EUR	35	33	36	38
Dividend per share	EUR	7.2	7.2	7.5	8.1
EPS/FFOper share	EUR	8	7.3	8.2	9

Figure 9.6 Typical analyst's format for forecasts.

Source: The authors.

9.4.4   Consensus forecasts

Aggregating these forecasts together will produce what is known as consensus forecasts. These can be accessed either via the buy side collating the analysts' reports it receives or via one of the third-party data and news sources such as Bloomberg or Refinitiv. Thus, when a company produces its figures, you will see commentary such as below, which relates to Prologis (PLD) Q4 2021 figures.

"PLD has reported fourth quarters 2021 core funds from operations (FFO) per share of US$1.12 beating the Zacks Consensus Estimate of US$1.10."

Source: Yahoo Finance

9.4.5   Inflexion points and direction of travel

Whilst a constant updating of models following figures is inevitable, the key point determining the reaction will be whether there is believed to be an inflexion point in the figures. By this, we mean an end to declining figures such as rents or NAV, which would be seen as good news, or indeed an end to growth in rents and values, which would be seen as bad news. The reason for this will be covered in the chapter on valuation but for now, it is important to understand that the trend or direction of travel of a company's key financial metrics will determine its valuation and subsequent share price performance.

9.5   Key institutions and data sources

Now that the importance of market structure, company forecasts, peer groups, index weightings, consensus estimates and market expectations have been explained, it is time to show the key information providers which support the activities of buy and sell-side participants. These can be grouped into trade associations and data providers.

9.5.1   Trade associations

Trade associations are vitally important to the real estate sector in terms of best practice, research, transparency, indices and liaising with policymakers, and as a platform for investors and asset owners to interact for the benefit of the sector as a whole. We show below the main participants, their website addresses, and their mission statements. We will be referring to their material throughout the book.

9.5.2 EPRA

Website: https://www.epra.com/

Mission statement (source: EPRA): "EPRA's mission is to promote, develop and represent the European public real estate sector. We achieve this through the provision of better information to investors and stakeholders, active involvement in the public and political debate, improvement of the general operating environment, promotion of best practices and the cohesion and strengthening of the industry."

9.5.3 NAREIT (National Association of Real Estate Investment Trusts)

Website: https://www.reit.com/nareit

Mission statement (source: Nareit): "Nareit serves as the worldwide representative voice for REITs and real estate companies with an interest in U.S. real estate. Nareit's members are REITs and other real estate companies throughout the world that own, operate, and finance income-producing real estate, as well as those firms and individuals who advise, study, and service those businesses."

9.5.4 APREA (Asia Pacific Real Assets Association)

Website: https://www.aprea.asia/

Mission statement (source: APREA): "APREA offers a distinctive global outreach and vision for the Asia Pacific real assets sector, with an aim to create value for our members through the following key foci:

Asia Pacific Opportunities
Professional Development
Real Assets Sector
Education & Research
Advocacy

9.5.5 INREV

Website: www.inrev.org

Mission statement: "INREV is the European Association for Investors in Non-Listed Real Estate Vehicles. We are Europe's leading platform for sharing knowledge on the non-listed (unlisted) real estate industry. Our goal is to improve transparency, professionalism, and best practices across the sector, making the asset class more accessible and attractive to investors."

9.5.6 Information providers

Most sell-side analysts will use an information provider for live share prices and historical share price information, company news and historical financials. The four most widely used platforms are:

Bloomberg
Refinitiv

FactSet
S&P (formerly SNL)

Quiz questions

1 Which market participants are behind the Chinese Wall?
2 List the key motivations of a sell-side analyst.
3 List the key motivations of a buy-side analyst.
4 Name the key variables required to assess a company's performance.
5 Why do investors use a free float market capitalisation weighted index?

Discussion questions

• How has the role of the sell-side analyst changed post-MiFID II?
• What impact does the number of analysts following a company have on its valuation?
• Why are consensus forecasts important?

Research/Dissertation topics

• Does index inclusion impact a company's performance and valuation?
• How accurate are analysts' forecasts and how often do they change?

Note

1 (Ref. https://corporatefinanceinstitute.com/resources/knowledge/finance/chinese-wall-definition/)

References

Accuracy of analysts' forecasts and recommendations and their importance in determining market capitalisation (firm value)

Devos, E., Ong, S.E. & Spieler, A.C. (2007). Analyst activity and firm value: evidence from the REIT sector. The Journal of Real Estate Finance and Economics, 35 (3): 333–356.
Freybote, J. & Carstens, R. (2021). Buy, sell or hold? The information in institutional real estate investor consensus. *Journal of Real Estate Research*, 43 (2): 181–203.

The impact of index inclusion on performance

Pavlov, A., Steiner, E. & Wachter, S. (2018). The consequences of REIT index membership for return patterns. *Real Estate Economics*, 46 (1): 210–250.

10 Analysing a public real estate company

Objectives

At the end of this chapter, you will understand:

- Why listed real estate is different from other equity market sectors.
- The five key accounting metrics that all analysts use.
- Difference between accounting metrics and investor-based metrics.
- Data sources to be used for forecasting.
- The role of EPRA NAV, EPS and cost ratio measures.
- How to undertake an investment analysis of any listed real estate company globally.

Key concepts

The difference and link between the Profit and Loss Account ("P&L") – aka revenue account or Income Statement – and Balance Sheet.

Difference between three types of metrics used:

- Per share data (used for valuation).
- Ratio analysis (used for peer group analysis).
- Qualitative analysis (used for valuation and investment recommendations).

Accounting (e.g., IFRS) data vs. restated best practice (e.g., EPRA) data.

10.1 Why listed real estate is different from other equity market sectors

There are 11 different sectors recognised in the equity market. Companies are classified into one of these sectors for benchmarking purposes. Until recently, real estate was included in the financial sector but in 2016 it became a separate sector. The mortgage REITs available in the US remain in the financial sector. So, what are the other sectors? They are:

- Information technology
- Health care
- Financials
- Consumer discretionary
- Communication Services
- Industrials
- Consumer staples

DOI: 10.1201/9781003298564-13

- Energy
- Utilities
- Materials

Within the real estate sector there are two classifications: REITs, and real estate management and development companies. As REITs comprise the most significant weighting globally (but not in every country), we focus on analysing them in this chapter whilst illustrating how to analyse non-REITs/Prop Cos.

It is important to understand the main conceptual differences between listed real estate and the other sectors. These can be narrowed down to the following:

10.1.1   Real estate is asset-based

The other sectors comprise predominantly trading companies which sell products or services. Since the advent of e-commerce and globalisation, this can be 24 hours a day, 7 days a week. As a result, the key items when analysing these companies are metrics such as gross and net profit margins, operating cost ratios, stock turnover and liquidity ratios. Real estate companies in contrast are either asset-based investment companies (REITs) or asset-based development companies with very lumpy sales. It should be noted that housebuilding companies are not included in the real estate sector, although certain Asian-based companies with large residential development arms are. As a result of being asset-based, standard metrics such as profit margins and liquidity ratios are not as relevant.

10.1.2   Income is less frequent but more predictable

Typically, a REIT receives its income four times a year (quarter days) from its tenant base (customers) with whom it has an ongoing contractual relationship (lease). This contrasts sharply with say, Apple, which sells a product every second of every day somewhere in the world. However, there is (normally) a greater certainty and predictability to the income of REITs due to the contractual nature of leases compared to the sales of a trading company. To put it at its most extreme, if the management team of a REIT did not turn up to the office for a period, the tenants would still be obliged to pay their rent. In contrast, if the employees and suppliers of a supermarket, say Sainsbury's, did not turn up to work, there would be empty shelves and no sales.

10.1.3   There is a parallel asset pricing market

All sectors have fundamental pricing in their underlying market, which influences the share prices of companies in that sector, be it the price of oil, financial products, electricity, baked beans, or semiconductors. However, real estate is different because the investment assets a real estate company owns are valued independently of the corporate entity (the listed company). If a REIT owns a shopping centre, that centre has an investment value and can be sold separately to the REIT.

10.1.4   Brand value

In contrast to real estate companies, the stock that other listed companies own, be it oil, electricity, baked beans, or semiconductors, is typically held at cost and requires an operational or distribution platform to realise full value. It is the listed entity that adds value by providing the operational platform and "brand". An unbranded Apple computer or pair of Adidas/Nike trainers is worth far less than the branded version. Typically, with the rare exception of say Westfield shopping centres, real estate assets have not been branded. We will return to this topic in detail in the chapter on valuation. It is worth noting at this stage however that the trend is changing. Not only do the "new", "alternative" sectors such as healthcare, data centres and student accommodation have significantly higher operational requirements, they also often have platforms which can add value, for example, Unite Students accommodation, to that of the underlying real estate.

10.2   The five key accounting metrics that analysts use

When looking at a listed real estate company or indeed an unlisted fund, five key factors need to be identified, each of which has an accounting metric we can use to identify how the company performs. Typically, regardless of geography, analysts will focus on the revenue account and balance sheet produced by companies. These can be extremely confusing for newcomers to the sector given the number of line items involved.

However, it is possible to simplify these as: 1) three items in the revenue account 2) and two in the balance sheet (Figures 10.1 and 10.2).

We can now look at these five items individually to understand what they represent and how they can be interpreted.

Revenue Account

10.2.1   Net rental income

This represents the top-line growth to the entity. It is the key driver of earnings and dividends, and a major contributor to capital values.

Financial Year to 31 March (£m)	2021	2022	Change %
Net rental income	367	429	16.9%
Fees & other income	11	13	18.2%
Administrative expenses	(74)	(89)	(20.3)%
Net finance costs	(103)	(102)	1.0%
Underlying Profit	201	251	24.9%
Underlying tax (charge) / credit[1]	(26)	4	
Underlying earnings per share (p)	18.8	27.4	45.7%
Dividend per share (p)	15.04	21.92	45.7%

FY22 dividend of 21.92p represent 80% of Underlying EPS

Figure 10.1 Revenue account example: British Land.

Source: British Land presentation – full year results 2023.

EPRA balance sheet

	31 March 2021	Group	Joint ventures	31 March 2022
Total properties (£m)[1]	9,140	6,938	3,538	10,476
Adjusted net debt (£m)	(2,938)	(2,495)	(963)	(3,458)
Other net liabilities (£m)	(152)	(177)	(70)	(247)
EPRA Net Tangible Assets (£m)	**6,050**	**4,266**	**2,505**	6,771
Loan to value (LTV)[2]	32.0%			32.9%
Weighted average interest rate[2]	2.9%			2.9%
Interest cover	3.0x			3.5x
Weighted average maturity of drawn debt (years)[2]	7.6			6.9

Figure 10.2 Balance sheet example: British Land.

Source: British Land presentation, full-year results 2023.

10.2.2   Administration expenses

Once we have established how much income is likely to be generated the next question is how much is it going to cost to run the portfolio and company? This is extremely important for several reasons: First, every pound/dollar/euro that is absorbed in administration expenses reduces the return to shareholders. Second, the relative cost of running a portfolio can be compared to the private markets (unlisted funds) and, third, it can be compared to other companies via the EPRA cost ratio. All companies disclose this information as administrative expenses, although it should be noted there can be inconsistencies between companies in terms of what is included. This also introduces us to the concept of internally managed (where the team are employees of the company) and externally managed (run by a third-party) REITs.

10.2.3   Net interest payable

The final item relates to the cost of servicing the debt of the entity/portfolio. Although overall leverage ratios have reduced post the GFC, virtually all listed real estate companies employ leverage. This has an impact on the capital value returns to shareholders but also, importantly, the revenue account. Typically, a company will have a portfolio of debt with varying terms and interest rates payable, so we need to understand the dynamics of the components rather than just looking at the average rate. Efficient treasury management can often yield benefits equal to if not greater than asset management. Similarly, at times of declining interest rates (for example, from 5.8% to 1.1% between 2008 and 2021), companies can refinance debt at a lower rate and generate a significant cash flow benefit (Figure 10.3).

10.2.4   Gross assets

Gross assets represent the underlying portfolio held by the entity and the key to capital returns. Particularly for those countries which incorporate revaluations into their balance

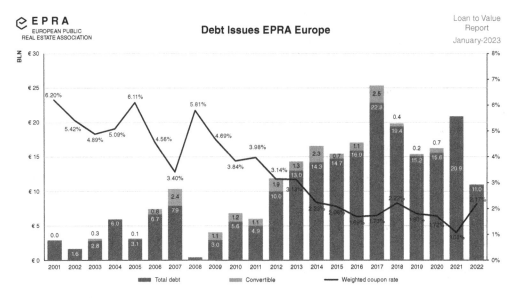

Figure 10.3 European debt issued by European listed real estate companies.

Source: EPRA.

sheet (UK and Europe), the key driver of shareholder returns will be capital returns, that is, the growth in the NAV, which is driven by the like-for-like change in portfolio values. It is important to understand that this excludes acquisitions/disposals and capital expenditure.

10.2.5   Net debt

The final piece of the jigsaw is to analyse the debt portfolio. This represents a management decision as to the optimal level of debt used to finance the portfolio. The impact of leverage is a major factor in determining investor level, as opposed to asset level, returns.

10.3   Difference between accounting-based metrics and investor-based metrics

The first look at a listed real estate company's financial information may be extremely confusing for newcomers to the sector (particularly in the US). Therefore, we often recommend, as a first step in analysing a company to go straight to the company's investors' presentations rather than the regulatory filings. These can normally be found on the company's website under the tab Investors, which then leads you to Investor Presentations. It is important to note that for analysis purposes, you should refer to full-year presentations, with interim and quarterly figures providing an update on the full-year figure.

Typically, within the investor presentation, a company will use "adjusted" or EPRA figures rather than IFRS or GAAP-based numbers. Why do they not use standard accounting metrics? Earlier in the chapter, we explained that the real estate sector was different from other sectors and the reasons for that. Standard accounting conventions do not always help interpret a company's financial statements. A good example is the treatment of revaluation surpluses/deficits. Under IFRS, these are shown in the revenue account. If a company's portfolio value increased by £1 billion over the accounting period, that £1 billion would be shown as "profit" even though it is not a realised trading figure but an unrealised capital surplus. Therefore, most companies in the real estate sector make some adjustments to the accounting figures to illustrate the true impact on their business. We recommend using EPRA best practice recommendations (BPRs) so that genuine comparisons can be made.

10.4   A sell-side approach to analysis

A sell-side analyst will typically follow between 10 and 20 companies. These might be all within one sector or one country or spread across a region.

For each of these companies, the analyst will build a detailed financial model. The purpose of this model is to enable the analyst to quantify very quickly the impact on leverage and the key per share metrics of NAV and EPS (and subsequently the DPS) of key events such as:

1 Company-related events –
  New lettings
  Disposals
  Acquisitions
  Equity issues
  Debt issues
  Bid approaches
2 Real estate market related events –
  Changes in estimated rental values.
  Changes in estimated investment yields
  Transactions on comparable assets –
3 Capital market events –
  Unexpected changes in interest rates, FX, inflation, market sentiment, geopolitical events, financial conditions

As such, every market change is factored into the model for those companies covered.

For buy-side analysts covering far more companies, it is not always feasible to undertake such a detailed task, so, often, they use external models and simplify them for internal purposes so that they can be comparable and allow sensitivity analysis.

A generalist fund manager will typically start by using consensus forecasts from a third-party provider such as Bloomberg or S&P and adapt them to their general equity model.

10.5   Data sources to be used

Now that we understand the structure of the financial model, we need to look at the different inputs. These are:

1 **Company data**
  Regulatory updates: Full-year figures, interim reports, quarterly updates
  Discretionary updates: Press releases on disposals, acquisitions, and market conditions
  Capital market days (site visits)
2 **Real estate market data**
  Real estate market indices
  Agents reports

3 **Capital market data.**
   Economic data such as GDP, inflation, employment, PSBR
   Changes to in-house forecasts

10.6   EPRA BPRs

EPRA produces best practice guidelines for companies that wish to report consistently for the following measures:

   Earnings
   Net asset value
   Net initial yield
   Vacancy rate
   Cost ratio
   LTV
   We recommend using these figures for consistency where they are available.

10.7   How to undertake the analysis of a real estate company

We now look at the five key items required to analyse a listed real estate company and provide details of the key questions to answer from the information available. This is shown under three headings:

1 Key issues to identify – this will provide an overview of the company's current profile in terms of quality, return profile and risk level.
2 Questions to ask – if market conditions change, is the company in a better or worse position?
3 Peer group scenario analysis – how is the company placed to respond to market changes relative to its peer group?

10.7.1   Rental income

Key issues to identify

- What is the weighted average unexpired lease term (WAULT).
- What is the reversionary profile – market or index-based reviews?
- Who are the key tenants?

Questions

- Under what market circumstances would it be beneficial to have many leases expire in the short term?
- How does the trade-off between tenant quality and rental income work?

Sample scenario analysis

- Which company in a peer group has the greatest potential for rental growth (ex-acquisitions)?
- Which company has the most exposure to a specific tenant (e.g., government, tech company, retailer) and what are the implications?

10.7.2   *Administration expenses*

Key issues to identify

- Is the portfolio internally or externally managed?
- How does the ratio of expenses to rental income compare to the peer group?
- Are there any transaction not performance-related incentives?

Questions

- To what extent are administration expenses controllable?
- Can large portfolios be run more efficiently?

Sample scenario analysis

- Which company provides the best and worst value for shareholders?
- What synergies could be extracted by a takeover?

10.7.3   *Interest payable*

Key issues to identify

- Relationship between disclosed interest payable and forecast levels of debt.
- Fixed/floating mix, weighted average cost.
- Relationship between rental income and interest payable (interest cover)

Questions

- Is it better to have fixed or floating-rate debt?
- What level of rental income/interest cover is acceptable?

Sample scenario analysis

- Which company in a peer group is most exposed to changes in interest rates?
- Which company would benefit most from a refinancing?

10.7.4   *Gross assets*

Key issues to identify

- Portfolio mix in terms of:
- Sector, location, tenant bias, yield profile
- Capital expenditure requirements, development exposure.
- Value-added opportunities

Questions

- What would your current preferred portfolio sectors and locations be?
- What would your preferred tenant profile be?

Sample scenario analysis

- Which company has the greatest exposure to specific locations, for example, the West End of London?

10.7.5   Net debt

Key issues to address

- Maturity/repayment profile
- Floating fixed rate split
- Secured/unsecured split.
- Quantum relative to gross assets
- Lenders
- Off-balance sheet debt

Questions

- What would your preferred level of gearing be?
- Is it fair to include off-balance sheet debt?

Sample scenario analysis

- Which company has the greatest exposure to specific lenders (e.g., Irish/Icelandic banks in GFC)?
- Which company has the greatest level of gearing?

British Land Company PLC					
INCOME STATEMENT (£M)					
Years ended 31 March	**Mar-10A**	**Mar-11**	**Mar-12**	**Mar-13**	**Mar-14**
Gross Rental Income	342	276	294	308	319
Property expenses	(5)	(6)	(7)	(8)	(9)
Net rental income	337	270	287	300	310
Operating expenses	(70)	(54)	(57)	(60)	(64)
Revaluation movements	514	191	130	141	243
Gains (losses) on disposals	(18)	-	-	-	-
Material one-off items	-	-	-	-	-
Income from associates	479	219	195	189	320
Other	-	-	-	-	-
EBIT	1,242	626	555	569	810
EBIT exc. valuation movements	728	435	425	429	567
Interest receivable	18	6	7	7	7
Interest payable	(158)	(80)	(90)	(112)	(134)
Capitalised interest	12	6	7	9	10
Net interest	(128)	(68)	(76)	(96)	(116)
Profit / (loss) before tax	1,114	558	479	473	693
Taxation	12	-	(5)	(5)	(7)
Profit after tax	1,126	558	474	468	686
Minority interests	-	-	-	-	-
Discontinued operations	-	-	-	-	-
Net income	1,126	558	474	468	686

Figure 10.4 Revenue account forecasts.

Source: Author, Macquarie.

10.8   Steps to produce forecasts for a listed real estate company

We now turn to the process of progressing the analysis to produce forecasts for the company. Assuming a good level of knowledge about the company's current position, the analyst will then use the following sources to help produce forecasts:

Company guidance
Consensus real estate market forecasts
Internal and external estimates of forward interest rates and FX rates
Peer group forecasts
Equity market consensus forecasts

These forecasts are then converted into per share figures for peer group comparison. We deal with these calculations in the next chapter.

The process is as follows (Figures 10.4 and 10.8):

BALANCE SHEET (£M)					
Years ended 31 March	Mar-10A	Mar-11	Mar-12	Mar-13	Mar-14
Intangible fixed assets	10	10	10	10	10
Investment property	4,126	4,405	4,936	5,463	6,092
Development property	-	22	123	219	316
Other	33	26	21	17	14
Tangible fixed assets	4,159	4,453	5,080	5,700	6,422
Non-current investments & other	1,855	1,999	2,116	2,227	2,468
Cash & Short-term investments	269	300	300	300	300
Debtors	85	69	73	77	79
Inventory	-	-	-	-	-
Other	20	20	20	20	20
Current assets	374	389	393	397	399
Total assets	6,398	6,850	7,599	8,333	9,300
Short term debt	139	-	-	-	-
Trade Creditors	104	10	12	14	16
Other	228	228	228	228	228
Current liabilities	471	238	240	242	244
Long term debt	1,642	1,991	2,488	2,974	3,475
Other	30	30	30	30	30
Provisions	47	47	47	47	47
Non-current liabilities	1,719	2,068	2,565	3,051	3,552
Share capital & premium	1,461	1,461	1,461	1,461	1,461
Reserves	2,747	3,083	3,333	3,579	4,043
Equity	4,208	4,544	4,794	5,040	5,504
Minority Interest		-	-	-	-
Total liabilities	6,398	6,850	7,599	8,333	9,300

Figure 10.5 Balance sheet forecasts.

Source: Author, Macquarie.

NAV FORECAST

	2010 £M	%	Mar-11 £M	£/share	Mar-12 %	£M	£/share	Mar-13 %	£M	£/share	Mar-14 %	£M	£/share
Retail warehouses	1,676.0	6.0%	100.6	0.12	2.0%	35.5	0.04	3.0%	54.4	0.06	5.2%	97.1	0.11
Superstores	166.0	3.0%	5.0	0.01	2.0%	3.4	0.00	2.0%	3.5	0.00	5.7%	10.1	0.01
Shopping centres	199.0	2.0%	4.0	0.00	2.0%	4.1	0.00	3.0%	6.2	0.01	5.6%	11.9	0.01
Department stores	436.0	1.0%	4.4	0.01	1.0%	4.4	0.01	1.0%	4.4	0.01	5.5%	24.7	0.03
High street	-	(2.0%)	-	-	0.0%	-	-	1.0%	-	-	5.5%	-	-
Total retail	**2,477.0**	**4.6%**	**113.9**	**0.13**	**1.8%**	**47.4**	**0.05**	**2.6%**	**68.5**	**0.08**	**5.3%**	**143.8**	**0.17**
City	493.0	5.0%	24.7	0.03	5.0%	25.9	0.03	4.0%	21.7	0.03	5.3%	30.0	0.03
West-end	967.0	5.0%	48.4	0.06	5.0%	50.8	0.06	4.0%	42.6	0.05	5.3%	58.8	0.07
Provincial	27.0	1.0%	0.3	0.00	1.0%	0.3	0.00	1.0%	0.3	0.00	4.9%	1.4	0.00
Total offices	**1,487.0**	**4.9%**	**73.3**	**0.08**	**4.9%**	**76.9**	**0.09**	**3.9%**	**64.7**	**0.07**	**5.3%**	**90.1**	**0.10**
Industrial / distribution / leisure / other	188.0	2.0%	3.8	0.00	3.0%	5.8	0.01	4.0%	7.9	0.01	4.9%	10.1	0.01
Total	**4,152.0**	**4.6%**	**190.9**	**0.22**	**3.0%**	**130.1**	**0.15**	**3.2%**	**141.1**	**0.16**	**5.3%**	**244.0**	**0.28**
Opening NNAV per share	**5.04**			**5.04**			**5.30**			**5.59**			**5.87**
Closing NNAV per share				**5.30**			**5.59**			**5.87**			**6.40**

Figure 10.6 Detailed NAV forecasts.

Source: Author, Macquarie.

LEVERAGE METRICS					
	Mar-10A	Mar-11	Mar-12	Mar-13	Mar-14
Net rental interest cover (x)	2.6 x	4.0 x	3.8 x	3.1 x	2.7 x
FFO	241	231	232	221	206
FOC	445	154	229	219	205
Net debt (£M)	(1,512)	(1,691)	(2,188)	(2,674)	(3,175)
Net debt / Equity (%)	35.9%	37.2%	45.6%	53.1%	57.7%
FFO / Total debt (%)	13.5%	11.6%	9.3%	7.4%	5.9%
FFO / Net debt	15.9%	13.7%	10.6%	8.3%	6.5%
Loan to value - gross (%)	43%	45%	49%	52%	54%
Proportional LTV	26%	26%	24%	28%	30%
Gross debt / Net tangible assets	42.4%	43.9%	52.0%	59.1%	63.3%

Figure 10.7 Derived leverage metrics.

Source: Author, Macquarie.

PER SHARE DATA					
	Mar-10A	Mar-11	Mar-12	Mar-13	Mar-14
Basic NAV (£ / share)	4.86	5.25	5.54	5.82	6.36
NNAV (£ / share)	5.04	5.30	5.59	5.87	6.40
Growth	32.4%	5.2%	5.5%	5.1%	9.1%
EPRA NAV	5.04	5.48	5.77	6.05	6.58
Dividend (£ / share)	0.26	0.26	0.26	0.26	0.26
Growth	(13.3)%	-	-	-	-
EPS (£)	1.33	0.65	0.55	0.55	0.80
Adjusted EPS	0.28	0.27	0.28	0.27	0.25
Growth	(31.0)%	(4.3)%	2.5%	(3.9)%	(4.9)%

Figure 10.8 Forecasts converted into per share figures.

Source: Author, Macquarie.

We have covered the process for analysing a listed real estate company and producing forecasts. In a subsequent chapter, we will look at how these are converted into per share metrics and how to value the company.

Quiz questions

1 Why is real estate different from the other ten equity sectors?
2 What are the five key accounting metrics used to value a listed real estate company?
3 Why would a company choose not to use an accounting definition of earnings?
4 What metrics do the six EPRA BRPs cover?
5 What are the three key sources of data used for forecasting?

Discussion questions

- How do the approaches of sell-side and buy-side analysts differ in producing company forecasts?
- What are the most pressing questions facing REITs at present?
- How accurate would you expect independent analysts' forecasts to be?

Research/Dissertation topics

- How do you determine alpha generation from a listed real estate company?

11 Analysing public real estate debt

Objectives

At the end of this chapter, you will understand:

- Key terms and characteristics of REIT bonds.
- How to measure the size of a REIT bond market.
- Trends in bond issuance.
- Credit rating methodologies and analytic processes.

Key concepts

- The difference between REIT bonds and CMBS.
- Key features of REIT bonds.
- Similarities and differences between real estate and other industries' bonds.
- Positive and negative covenants.
- Relationship between the cost of debt and issuance.

11.1 Characteristics of REIT bonds

It is important to distinguish at the outset between REIT bonds and mortgage REITs (which are prevalent and popular in the US but not elsewhere). Mortgage REITs do not invest directly in real estate, but in securities that gain exposure to the real estate market. This could be either directly via holdings of REIT bonds or indirectly by securitised packages of loans, such as mortgage-backed securities. Crucially, this means that, in the event of a bankruptcy, these instruments rank lower than REIT bonds. Additionally, mortgage REITs are subject to fewer legal restrictions and can have complex holdings, so it is more difficult for investors to analyse the risk profile compared to REIT bonds.

It is also important to remember the distinction made in previous chapters between REITs and developers. The fundamental structure of REITs (identifiable stabilised income-producing assets held in a tax-efficient manner) lends itself to public debt (bond) issuance, whilst the nature of developers, where most returns are derived from capital rather than income, cashflows are irregular and more unpredictable, with a greater immediate impact on values from economic uncertainty, means that their bonds have a different risk profile and investor appetite.

For each bond, we can identify the following key features, which determine investor appetite for the issue. This list also provides an overview of key terms used in the analysis and the valuation of public real estate debt.

DOI: 10.1201/9781003298564-14

- **Coupon** (interest rate) – Fixed rate or variable. Occasionally bonds do not have a coupon but are issued below par value so the return to investors is the difference between the issue price and redemption at par. Normally, however, most of the return to investors is via the coupon or interest payable on the bond.
- **Term** – Fixed term or perpetual. Although the idea of a perpetual bond appears initially attractive, the structural changes in the real estate market and indeed the trend towards shorter leases means that better pricing is available for fixed-term bonds, where the outlook has more visibility. When leases in the UK were 25 years, it was common to have 25-year debentures issued, but now, typically, a bond will be issued with a term of 3–10 years.
- **Ranking** – Senior or subordinated. This relates to the order of payout in the event of bankruptcy. Senior debt takes precedence and therefore attracts more favourable pricing. Bond investors will look in detail at the capital stack of a REIT to determine their priority in terms of payout.
- **Size** – The size of the issue determines liquidity and investor appeal. It should be noted that whilst there are specialist bond fund managers who trade their portfolios, a significant portion of investors will seek to hold the bond issue to maturity when it is repaid at par. This is particularly true of investors with asset-liability matching strategies in place. Therefore, the relationship between size and liquidity is not the same as for equities, but it is true that the greater the size of issue, the broader the investor base that can participate.
- **Issue Price** – If the bond is issued at below par value, there will be some capital appreciation when the bond is repaid. Pricing of the bond will reflect market conditions at the date of issue.
- **Yield-to-Maturity** – This reflects the total return to the investor throughout the life of the bond and is used in secondary market valuations when bond prices differ from the issue price. This calculation allows investors to benchmark bonds against each other and an index.
- **Covenants** – financial constraints the issuer must abide by or provide a remedy if they are breached. These are designed to protect both the issuer and the investor. They can be either positive (the issuer benefits if they achieve certain targets) or negative (the issuer must remedy any breach so that it remains compliant with the covenants). Examples of covenants include:

  - Staying within specific financial ratios, for example, LTV and DSCR.
  - Maintaining best practice accounting standards and reporting (either IFRS, GAAP or EPRA BPRs).
  - A restriction on selling specific assets
  - Not taking on additional debt
  - Not taking on debt which ranks higher (i.e., super senior) to the bond
  - Materially change the nature of its business

- **Secured or unsecured** – Secured bonds are secured against specific assets whilst unsecured are effectively a corporate loan. Typically, unsecured bonds are only available to larger companies with an established track record.
- **Currency** – This can match either the asset base or the investor base. Investors group bond issuances by currency.
- **Issuer** – This can be either the REIT parent company or a subsidiary/joint venture.
- **Sustainability criteria** – Those with sustainability-linked covenants are known as green bonds and have a wider potential investor pool. This is becoming increasingly popular with investors.

- **Pure debt or hybrid** – Certain issues have an equity component in addition to the plain vanilla debt. This is known as a convertible bond, and whilst the coupon may typically be lower than for a straightforward bond, the number of investors who participate in convertibles is smaller than for straight bonds. The coupon is lower because there is theoretically some upside via the conversion rights to equity. The conversion price will be at a premium to the share price at the time of the convertible issuance.
- **Credit rating** – Typically issuers pay credit rating agencies to have their bonds analysed and rated. This informs the issuer of the risk level of the bond. Rating agencies have a different nomenclature but, typically, we can rate bonds as:
  Highest quality = AAA
  Investment grade = BBB and above
  Below investment grade, a.k.a. junk bonds = BBB and below
  Higher risk of default = C
  In default = D

11.2   The size of the market

11.2.1   Duration

When looking at bond markets, we typically approach it in three ways. First, we look at the market in terms of bond duration, second, in terms of credit rating and, finally, in terms of the underlying real estate sector. In the following tables, we use data from the real estate bond market (i.e., REITs and REOCs) for Europe and the UK to show the choices available to investors. All the data for these charts is from the Bayes Quarterly Bond Research publication, Q3 2022 (Figure 11.1).

First, we segment the market by duration. From an investor's point of view, it is important to note not just the size of the market for each duration but also the number of issues (i.e., from different companies). It should be noted that some companies have multiple issues, so, for example, in Europe, there are not 396 different companies with a bond with a maturity under three years, but 396 separate bonds listed. Two things are

EUROPE				
Years to Maturit	No of Bonds	Total Amount Out. (in million)	Coupon (weighted)	YTM (weighted)
0--3	396	64,896	2.1%	22.6%
3--5	267	70,211	2.0%	6.7%
5--7	143	44,071	1.3%	6.2%
7--10	128	44,503	1.4%	5.8%
10--15	71	15,932	1.5%	5.4%
15--20	46	4,187	1.7%	5.5%
20--30	42	3,205	2.1%	4.5%
30+	5	29	1.9%	3.9%
TOTAL	1098	247034		

UK				
Years to Maturit	No of obs	Total Amount Out. (in million)	Coupon (weighted)	YTM (weighted)
0--3	32	5,232	3.4%	7.6%
3--5	36	8,696	3.1%	7.5%
5--7	40	10,151	3.4%	7.9%
7--10	31	8,703	3.4%	7.1%
10--15	36	9,643	3.5%	6.7%
15--20	40	10,217	5.2%	6.9%
20--30	84	22,929	3.7%	0.3%
30+	26	5,721	2.6%	5.5%
	325	81292		

Figure 11.1 Bond duration.

Source: Bayes Business School.

EUROPE				
Rating group	No of Bonds	Total Amount Out. (in million)	Coupon (weighted)	YTM (weighted)
AAA	0	-	-	-
AA	0	-	-	-
A	109	36,024	1.2%	4.6%
BBB	135	66,412	1.5%	5.7%
BB	14	4,745	3.2%	10.7%
B	4	1,740	3.1%	11.8%
C	33	12,144	1.7%	11.2%
TOTAL	295	121065		

UK				
Rating group	No of Bonds	Total Amount Out. (in million)	Coupon (weighted)	YTM (weighted)
AAA	2	544	4.8%	6.2%
AA	9	2,906	3.2%	6.2%
A	105	27,100	3.5%	6.5%
BBB	46	15,958	4.1%	7.8%
BB	7	2,720	3.3%	10.4%
B	0	-	-	-
C	10	2,486	4.4%	6.5%
	179	51714		

Figure 11.2 Credit rating.

Source: Bayes Business School.

noticeable from this table. First, for the European short-maturity bonds (0–3 years), the yield-to-maturity is extremely high, at 22.6%. This represents how, for certain bonds, the current market pricing suggests there is a risk they may not all be repaid in full. Second, the European market is approximately three times the size of the UK market by overall volume and by number. Third, reflecting the trend in government bond yields and UK and ECB interest rates, the weighted average yield-to-maturity for European bonds is lower than for the UK (Figure 11.2).

11.2.2 Credit rating

We now turn to the second way of analysing the size and depth of a market by looking at the amount and size of bonds in each credit rating band split between Europe and the UK.

The first point to note is that most bonds issued are in the A and BBB categories. In other words, they are investment grade but are not triple A rated. This reflects, *inter alia*, the size of the companies, as real estate is a relatively small sector in market capitalisation terms. The second point to note is that as one would expect, the yield-to-maturity rises as the credit rating decreases. It is important to note that it is key to use the yield-to-maturity figure, as this represents market pricing, and not the coupon figure, as that will reflect the vintage, that is, the prevailing interest rates at the time of issue.

11.2.3 Sector exposure

Finally, in terms of investor choice, it is important to understand the underlying sectors to which the bonds provide exposure. In Figure 11.3, we show the breakdown for the UK and Europe by property type.

As the market evolves, we would expect to see far less concentration on shopping centres and more on new emergent sectors, such as healthcare, and all forms of residential.

11.3 Trends in issuance

There are several factors which influence the level of issuance of REIT bonds throughout the cycle. These include the cost and availability of debt, the outlook for interest rates,

EUROPE					UK			
Industry	No of Bonds	Total Amount Out. (in million)	Coupon (weighted)		Industry	No of Bonds	Total Amount Out. (in million)	Coupon (weighted)
health	29	1,891	1.2%		health	7	1,993	3.5%
leisure	4	242	3.2%		leisure	8	1,317	4.7%
logistic	69	30,397	1.5%		logistic	13	3,607	2.3%
office	48	11,670	1.6%		office	34	8,660	3.5%
retail	72	25,585	1.7%		retail	20	5,424	3.8%
supermarket	0	-	0.0%		supermarket	6	3,203	5.9%
resi	194	59,802	1.4%		resi	159	40,376	3.6%
student	14	3,089	1.6%		student	14	2,060	3.4%
diversified	210	43,468	1.7%		diversified	15	1,625	4.2%
commercial dev	118	6,783	3.6%		commercial dev	0	-	
data centre	2	1,100	0.7%		data centre	0	-	
resi dev	64	11,561	1.7%		resi dev	0	-	
senior living	9	667	1.8%		senior living	0	-	
hotel	18	1,554	2.3%		hotel	0	-	
storage	10	1,599	1.0%		storage	0	-	
shopping centre	33	10,723	2.0%		shopping centre	0	-	
social resi	5	210	2.5%		social resi	0	-	

Figure 11.3 Sector exposure.

Source: Bayes Business School.

equity market valuations, investor appetite for leveraged companies and global liquidity. First, let us examine how the level of issuance has changed over the last 20 years.

As can be seen from Figure 11.4, one of the features of real estate capital markets post the GFC has been the significant level of issuance of debt relative to equity issuance. It is

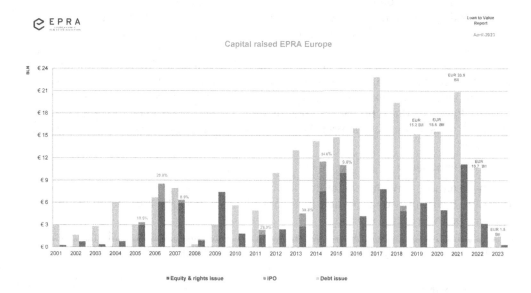

Figure 11.4 Bond vs. equity fundraising Europe 2021–2022.

Source: EPRA.

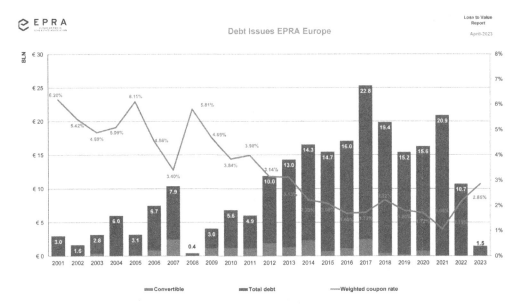

Figure 11.5 Cost of debt and issuance.

important to note that not all of this necessarily represents additional debt; but a significant amount can be seen as refinancing existing debt.

The level of debt issuance is going to be driven by the cost of debt and investor appetite. Figure 11.5 shows how the level of issuance rises as the cost of debt falls. 2022 saw a turning point in the interest rate cycle, and going forward, we would expect the level of debt issuance to be reduced if interest rates remain at elevated levels.

Looking at the US, the picture is not dissimilar, although it should be noted that the interest cycle is different from the UK and Europe (Figure 11.6).

Perhaps the most important development has been the increasing use of green bonds. Figure 11.7 shows how this has rapidly gained pace since 2018 and accounted for circa 18% of all issuances in US REITS. We would expect this figure to continue to rise significantly in line with investor appetite for higher levels of sustainability disclosure and compliance.

11.4 Credit rating methodology

Finally, we need to look at how credit rating agencies analyse a bond issue from a real estate company. Each rating agency has different methodologies, but we can provide an outline of the key points to cover similar to the approach we took with analysing the equity of a REIT earlier in the book. We can split the analysis into the following sections:

- Country analysis
- Sector analysis
- Portfolio quality
- Management quality
- Financial profile
- Peer group analysis

Total US REIT bond issuance by year

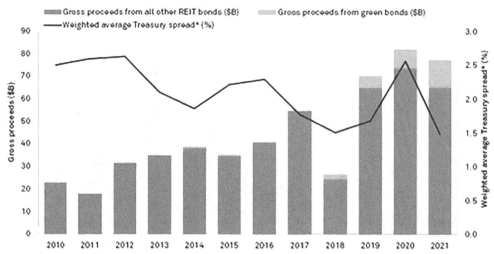

Data compiled Jan. 28, 2022.
Includes senior debt bond issuances by S&P Global Market Intelligence-covered equity REITs completed between Jan. 1, 2010, and Dec. 31, 2021.
* Represents the difference in yield between the underlying debt security and a U.S. Treasury note with the same maturity date at the time of issuance.
Source: S&P Global Market Intelligence

Figure 11.6 US REIT bond market.

Source: S&P Global Market Intelligence.

US REIT green bond issuance by year

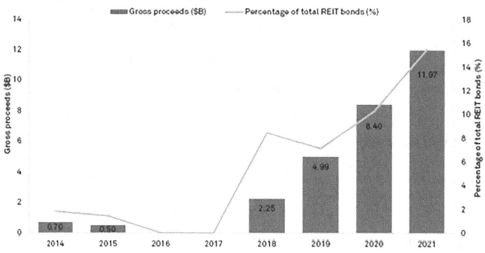

Data compiled Jan. 28, 2022.
Includes senior debt bond issuances by S&P Global Market Intelligence-covered equity REITs completed between Jan. 1, 2014, and Dec. 31, 2021, with an intended use of proceeds related primarily to green projects.
Source: S&P Global Market Intelligence

Figure 11.7 US REIT bond market – green bonds.

We can now look at a specific example, Morningstar, and how they structure their approach to credit rating US REITs. Their methodology is a sector-specific model, and the REIT methodology is based on the same four key components as for non-financial corporations, namely:

- Business risk
- Cashflow cushion (proprietary to Morningstar)
- Solvency score (proprietary to Morningstar)
- Distance to default

They adjust the first two elements to account for the differences in capital structure and the business model between REITs and other non-financial corporations. Their model combines qualitative judgements with quantitative financial and market data to arrive at a model-derived credit score. However, the model score is only an input to the final rating decision, which may consider trends in performance, anticipated company actions, macroeconomic and other developments that may not be reflected in the model.

This can also be shown in diagrammatic form. Moody's for example use the following framework to produce their scorecard (Figure 11.8).

Finally, it is instructive to use an example from a credit rating report on a REIT to understand the language and purpose of their reports. The following is an extract from a Fitch report on SEGRO from September 2022.

Illustration of the REITs and other commercial real estate firms methodology framework

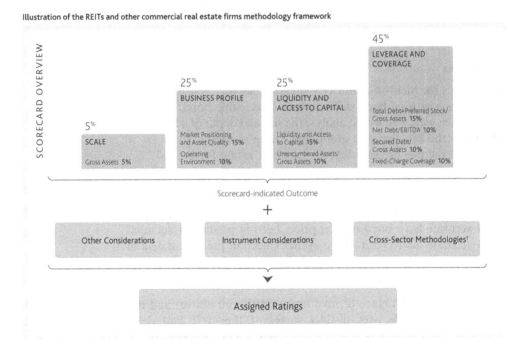

Figure 11.8 Moody's framework.

Source: Moody's.

Example: Fitch report on SEGRO, September 2022

"SEGRO PLC's ratings reflect the quality of its portfolio of urban warehouses and big-box assets located in the UK and the rest of Europe. The portfolio has continued to benefit from high tenant demand, because of increasing e-commerce penetration, and a constrained supply of new logistics assets.

This has supported the portfolio's high occupancy rate of 96.8% and like-for-like net rental income growth of 4.9% in 2021. Fitch Ratings expects net debt/EBITDA to decline as SEGRO's portfolio generates further rental growth and raises equity to fund its development and acquisitions.

Leverage to Decline: We expect SEGRO's net debt/EBITDA (using annualised rental income) to decline to 8.4x in 2022 and 7.9x in 2023, from 8.7x in 2021. The Fitch-adjusted LTV, including its JV SELP at share) remained stable at 25% at end-2021 (2020: 26%), aided by like-for-like valuation gains of 29% for the group's portfolio. We expect EBITDA/net interest expense to be about 6x in 2022 and 2023.

Rental Growth to Continue: Fitch expects SEGRO's portfolio to generate rental growth, particularly within the London and Paris urban warehouse portfolios, where scarcity of land and competing land uses restrict supply and tenant demand is high. For big-box warehouses in the UK and continental Europe, where there is greater demand, we expect rental growth to at least keep pace with inflation. In the UK, particularly in the urban portfolio, rental growth can be captured at rent reviews and lease renewals."

As can be seen, the tenor and tone are not dissimilar to that of an investment analyst's report. The key difference is that whereas an equity analyst's report is focused on the potential change in market valuation to produce a recommendation, the credit rating agencies focus on creditworthiness and the likelihood of default.

Quiz questions

1 What does the yield-to-maturity represent and why is it used?
2 How does a normal bond differ from a convertible bond?
3 Name three ways you would analyse the size and depth of a bond market for an investor,
4 What are the key elements of a credit rating methodology?
5 What are the five broad categories of credit rating for bonds?

Discussion questions

• How do the approaches of a credit rating analyst and an equity analyst differ?
• How accurate have bond market prices been in predicting defaults?
• How accurate have credit rating agencies been in predicting defaults?

Research/Dissertation topics

• Examine the performance of investment (BBB and above) and non-investment grade (below BBB) bonds over distinct stages of the cycle and determine whether the perceived higher risk has generated a higher level of risk-adjusted return.

12 Valuing a public real estate company

Objectives

At the end of this chapter, you will understand:

- How to value any listed real estate company globally.
- How to identify and apply key per share metrics.
- Defining the per share metrics.
- Three different valuation methodologies.
- Regional variations in valuation methodology and the reason for them.
- The sensitivity of valuation shifts.
- The time-variant nature of "fair value".

Key concepts

- Any company which has a share price will have a current valuation.
- Different valuation methodologies can be applied and are equally valid according to the investor type.
- The impact of short-term equity market movements on valuations.
- Deriving an implicit forecast at times of market turbulence.

12.1 How to value any listed real estate company globally

In an earlier chapter, we took that first step (company analysis) in the process of producing a valuation, target price and investment recommendation for a listed real estate company. We now need to build on that analysis to produce a valuation. In this chapter, we will:

- Calculate the key per share metrics.
- Apply per share forecasts of the key five accounting metrics we looked in a previous chapter.
- Calculate the three valuation metrics and determine an appropriate one.

In this chapter, we will take the final step and determine a target price and an investment recommendation.

12.2 Identifying and defining key per share items

To produce a valid valuation model for a listed real estate company, we need to identify and understand the key metrics which accurately measure the performance of the company. We also need to understand which valuation models are the most dynamic, that is, where they

DOI: 10.1201/9781003298564-15

Financial highlights

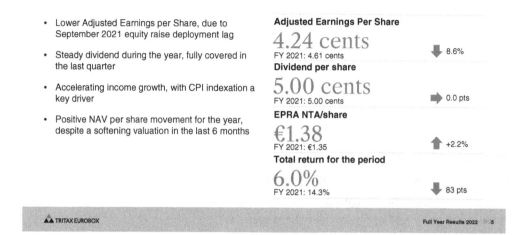

- Lower Adjusted Earnings per Share, due to September 2021 equity raise deployment lag

- Steady dividend during the year, fully covered in the last quarter

- Accelerating income growth, with CPI indexation a key driver

- Positive NAV per share movement for the year, despite a softening valuation in the last 6 months

Adjusted Earnings Per Share

4.24 cents
FY 2021: 4.61 cents

⬇ 8.6%

Dividend per share

5.00 cents
FY 2021: 5.00 cents

➡ 0.0 pts

EPRA NTA/share

€1.38
FY 2021: €1.35

⬆ +2.2%

Total return for the period

6.0%
FY 2021: 14.3%

⬇ 83 pts

▲▲ TRITAX EUROBOX   Full Year Results 2022   P 5

Figure 12.1 Tritax Eurobox metrics.

Source: Tritax Eurobox.

can adjust very quickly to either changes in the share price or changes in our or the market's assumptions.

We have previously identified five key accounting metrics that we need to understand to determine how a company is performing. We now adopt a similar approach to determining the key valuation items. These should all be capable of being expressed on a per share basis so that they can be comparable. The three key performance measures and per share metrics we can identify and that satisfy our requirement of accurately measuring the company's financial performance are:

1 **Cashflow generation = Earnings per share (EPS) "Also known as Funds From Operations" (FFO) per share in the US.**
   How much is the company expected to generate, after tax, on an annual basis?
2 **Cash on cash return to investors = Dividend per share (DPS)**
   How much can shareholders expect to receive in terms of annual cash income return?
3 **Value of the underlying assets = Net Asset Value per share (NAV)**
   What is the market value of the underlying assets (less associated debt and other liabilities) regardless of share price?

These three metrics provide a comprehensive snapshot of the company's financial performance. An example of how this is presented to shareholders is shown in Figure 12.1.

12.3   Defining the per share metrics

We now need to define each of these three per share metrics to ensure consistency when comparing companies.

1 EPS and FFO

Definition:
Earnings attributable to ordinary shareholders/weighted average number of shares during the accounting period = EPS.
Comment
It is always best to use company stated "Adjusted" EPS OR EPRA EPS or FFO, which smoothes out accounting inconsistencies.

2 DPS

Definition:
DPS is the amount declared by the company. It is not an accounting measure and is determined by the board.
Comment
The dividends are normally split into interim (paid after six months) and final (paid at the full year) to make the total DPS. There can also be special dividends, but these are not included when calculating dividend yield. They are included when calculating total shareholder return.

If shares qualify for the dividend, they are traded cum-dividend; if they are not entitled to the dividend, they are ex-dividend.

3 NAV

Definition:
Net Asset Value Per Share (NAV) = Net assets at the end of the accounting period/shares in issue at the end of the accounting period.
Comment
It is always best to take the most conservative figure for valuation purposes. NAV (along with EPS) can be shown as basic or fully diluted (after converting warrants, options, and convertible bonds to equity). Typically, the fully diluted figure is lower than the basic figure.

Always use EPRA NAV figures if they are available.

12.4 Three key valuation methodologies

In an earlier chapter, we discussed how listed real estate was different from other sectors. The valuation methodologies we choose should therefore seek to highlight the characteristics of the asset class. We first look at why we need real estate equity valuations; then we look at the valuation models available before determining the most appropriate for the real estate sector.

12.4.1 *What do we use valuations for?*

There are three reasons for using a real estate equity valuation.

1 Comparing companies, a.k.a. peer group analysis.
   To determine the appropriate valuation for a listed company, we need to be able to consider how the valuation compares to other companies with similar assets, investment strategy, size, shareholder base and leverage.

2 Comparing stock market valuations to direct market valuations – parallel asset pricing. A key differentiator for the real estate sector is, as we have discussed, the pricing of the underlying sector (real estate assets). We can compare how the equity market values the entity holding these assets with the underlying real estate valuation.

3 Calculating implied growth rates – implied valuations.

At times of market turbulence, we can take the current valuation, and compare it to the long-term average to determine an implied level of growth (or decline). We then compare this rate with our explicit forecasts to determine an appropriate valuation.

12.4.2   Valuation models available

A simple glance at either a finance textbook or a broker's note on a listed company will show that we have a (potentially bewildering) number of methodologies open to us to use. These include:

DCF models
Dividend discount models
Price earnings multiples
Sum of the parts valuation
Free cashflow yield
Discount/premium to NAV
Discount/premium to book value
Implied yield models
EBITDA/EV Models
Economic value-added models
Cost-of-capital models

12.4.3   Appropriate for the real estate sector

To simplify and make the valuation relevant to real estate, we can say that there are three potential valuations methodologies.

1 Earnings-based models
2 Dividend-based models
3 Asset-based models

We now go through an example to show how these models can easily be applied.

Example: Reliable REIT – Share price 200p, est. NAV in one year is 250p, estimated earnings or AFFO are 12p per share, and est. dividend is 10p per share. Dividends are expected to grow by 3% pa and the LTV is 50%.

10-year bond yields are 2% and investors' required rate of return on REITs is 8%.

12.4.4   What are the current valuation metrics of the company?

1 EARNINGS METHOD

Methodology: Share price/forecast company EPS.

Valuation = 200p/12p = **16.7 × earnings or AFFO**, or **earnings yield** of 1/16.7, × 100 = **6%.**

Rationale: A multiple of the company's estimated post-tax earnings capitalises the estimated cashflows attributable to the corporate entity – this method therefore is **valuation of the company**.

Advantages
Enables comparison with other equity sectors.
Can be expressed either as a multiple or as a yield.
Reflects cashflow projections of a company.

Disadvantages
Distorted by IFRS.
Less relevant for an asset-based industry – designed for trading companies with operating margins.
Companies provide adjusted EPS figures, which can use different assumptions.

2 NAV METHOD

Methodology: Share price/forecast NAV per share.
Valuation = 200p/250p = 0.8, then –1 – 0.2 and x 100 = **discount to NAV of 20%.**
Rationale: the premium or discount to forecast NAV captures expected growth and allows a **valuation of the company's assets**, and comparison with pricing in the direct real estate market.
Used to compare public markets with Private Markets.

Advantages
Suitable for asset-based industry where the key driver of growth is an uplift in values.
Enables comparison with private markets and underlying real estate.
Using forecast NAVs provides a level of implied growth in values.

Disadvantages
Distorted by level of the company's gearing.
Most suited to fully let investment companies – they can understand the worth of developments if they are not marked to market.
Can reverse quickly, and it is debatable how much companies have an influence (alpha), or if it is market-driven.

3 DIVIDEND MODELS

Methodology: Dividend/share price = dividend yield.
Valuation: 10p/200p = 0.05, × 100 = **dividend yield of 5%** compared to 10-year bond yields of 2% (a 300 Bp premium) and equity market yield of 3.5% (a 150 Bp premium).
Rationale: Dividend models allow direct comparisons of the expected cashflows that an investor will receive and is therefore a **valuation of the investor's income stream**.

Gordon growth model

$E = DIV / r\text{-}g$. Where E = value of shares. DIV = DPS, r = required rate of total return on equity, and g = constant rate of dividend growth.

Thus, if a company's dividend was 10p, r was 8%, then g is 3%, and E would be: 10/0.08–0.03 =10/0.05 = 200p.

Advantages

Measures true cash return to investors.

Allows direct comparison with return on underlying assets.

Can be used to compare to any capital (debt or equity) market globally.

Disadvantages

Best suited to REITs.

Highlights correlation with bond markets.

Understates the value of development companies.

Assumes a constant rate of dividend growth which in practice is unlikely

12.5 Regional variations

Of the three models available, the most used in Europe, where assets are revalued, are the dividend or distribution yield, and the premium/discount to NAV methods. We can combine these two valuations to produce a snapshot of where companies are trading on a global basis (see Figure 8.2) (Figure 12.2).

There are regional variations in terms of preferred valuation methodologies. These are:

- UK and Europe – mark-to-market assets, so NAV- and dividend yield-based methods favoured.
- US – no mark-to-market of assets, so multiple of AFFO and dividend yield key metrics.
- Australia – free cashflow yield and sum of parts methodologies favoured.

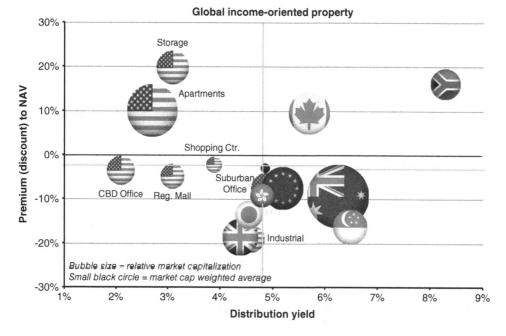

Figure 12.2 Global valuation comparison.

Source: Consilia Capital.

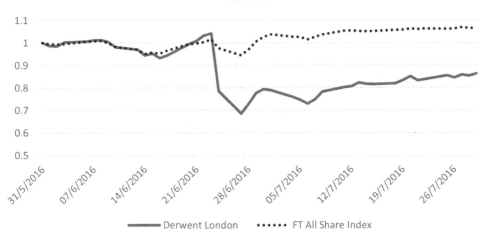

Figure 12.3 Immediate impact of EU Referendum.

Source: The authors, Bloomberg.

- Asia – market is development- and residential-based, so DCF NAV is common.
- At times of uncertainty over future capital value movements, investors have increasingly focused on dividend yield, highlighting the benefits of the tax-friendly REIT structure.

12.6 Sensitivity of valuation shifts

An important concept to understand is how quickly equity valuations can shift. We will deal with this in more detail in the chapter on target prices, but an example will help illustrate the point. Derwent London is a highly regarded central London office-based REIT with no financial problems and an exemplary 30-year track record. Prior to the EU Referendum vote in June 2016, it was highly valued. The morning after the unexpected result to leave the EU, the shares fell by 30% and did not immediately recover (see Figure 12.3).

Why?
The answer lies in investors' expectations. London offices were expected to be negatively impacted by the vote as financial services companies relocated to Europe. This did not occur at anything like the level assumed, and, indeed, overseas investment increased due to the decline in sterling. However, immediately after the vote, the equity market (not property) valuation of Derwent London changed immediately to reflect a lack of buyers and a lack of positive news in the foreseeable future.

12.7 Time-variant nature of fair value

The final point to consider is how the equity market valuations move over time. The easiest way to consider this is to look at the premium/discount to NAV valuations of the leading European countries post-GFC.

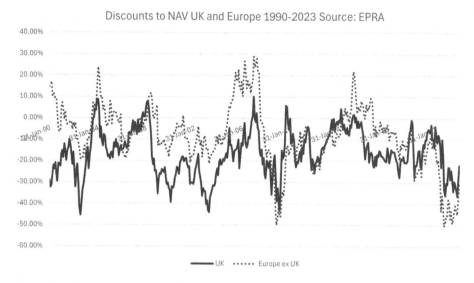

Figure 12.4 European valuation dispersion.

Source: EPRA.

As can be seen in Figure 12.4, the range of valuations over that period extends from a circa 30% premium to NAV to a circa 70% discount. This volatility reflects a continuous pricing mechanism, changing investor expectations and the price of liquidity. It is worth noting the influence of the general investors who are the key price setters rather than the specialist real estate investors. Real estate equities can therefore be seen as having an anchor to the NAV but have the ability to decouple from that anchor over time.

Quiz questions

1 What are the three performance metrics we look at for a real estate company?
2 Define EPS.
3 Define NAV per share.
4 What valuation method is prevalent in the US?
5 What are the advantages of the dividend valuation methodology.

Discussion questions

• How would you determine the appropriate valuation methodology for a company?
• How would a general investor differ in approach from a specialist real estate investor?
• Would a sell-side and buy-side analyst adopt similar valuation methodologies?

Research/Dissertation topics

• How do you determine if an equity valuation accurately reflects the growth prospects of a company?

13   Setting a target price

Objectives

At the end of this chapter, you will understand:

- The purpose of a target price.
- The range of different target price methodologies available.
- Difference between sell-side and buy-side approaches.
- Assumptions and catalysts that drive the target price.
- Sensitivity of target prices and how to adjust them in response to market conditions.
- How to make a logical investment recommendation.

Key concepts

- The current share price produces the current market valuation.
- A target price represents the expected share price at a specific point (1–2 years) in the future.
- Short-term equity market movements have a significant impact on valuations and implicit forecasts.
- The target price methodology should be consistent with the user requirements.

13.1   The purpose of a target price

A target price is set to determine the expected performance of a company's share price, based on detailed forecasts and allow comparison of that expected performance in absolute or relative terms, and produce a logical investment recommendation.

Step 1 is to determine the initial market valuation provided by the current share price and historical and either consensus or one's own forecast per share metrics. This is shown in Figure 13.1, which shows how a typical sell-side note would be laid out.

Normally for a listed real estate company there will be three different per share metrics shown (EPS, DPS, and NAV per share) and therefore three different valuation methods used (earnings, dividend and asset-based). In addition, these will be shown over three periods: historical (A), forecast year 1 (FY1) and forecast year 2 (FY2). For a sell-side analyst, the forecasts are their own, based on the detailed financial model and assumptions regarding the five key metrics discussed previously: Rental Income, Admin expenses, Interest Payable, Gross Assets and Debt. Buy-side analysts might use their own detailed forecasts, but more commonly, they would use consensus forecasts, that is, the average of all the brokers' forecasts. The reason for this is not just time. For a buy-side analyst,

DOI: 10.1201/9781003298564-16

Share Price	100

VALUATION METHOD	VALUATION Historic	FY1	FY2
Per Share Data			
EARNINGS EPS/AFFO	6	7	8
P/E or AFFO Multiple	16.67	14.29	12.50
DIVIDEND DPS	5	6	7
Dividend Yield	5.00%	6.00%	7.00%
NAV NAV	120	130	140
Discount/Premium to NAV	-16.67	-23.08	-28.57

Figure 13.1 Current valuation template.

Source: The authors.

understanding what the market is expecting from the company is crucial. Sell-side analysts are often looking to stand out from the crowd, as well as generate revenue by allowing their models to be used by the buy side, so detailed forecasts are required.

Step 2 is to apply the target price methodology, to these valuations and forecasts, which we will discuss in the next section. This is to produce a single figure to be compared to the current share price. This forms the basis for all commentary around the company and the logical foundation of the investment recommendation. If, for example, a company sells a building above expectations and market valuation, the analyst might upgrade the forecast for the portfolio values and change their target price accordingly.

13.2   The range of different target price methodologies available

The specific target price chosen by sell-side analysts will vary. We show four different examples here:

EXAMPLE 1

This involves a long-term average discount to NAV, which is then adjusted for specific factors to arrive at a fair value discount, which is applied to forecast NAVs. It is logical and simple to use and adjust.

EXAMPLE 2

This methodology looks at whether a company creates positive or negative returns (i.e., above or below its weighted average cost-of-capital) and subsequently accords a premium/discount valuation to forecast NAV. It is logical but slightly more complicated to determine.

EXAMPLE 3

The first two methodologies (all European-based) are very much focused on NAV as the key driver of future share price performance and therefore determining the appropriate

Method	Target Price	Weighting	TOTAL
Earnings based Target Price	126	0.33	41.58
Dividend based Target Price p	150	0.33	49.50
NAV Based Target Price	111	0.34	37.74
WEIGHTED AVERAGE Total		1	128.82

Figure 13.2 Weighted average methodology.

Source: The authors.

discount or premium. In this last example (devised by the author while at Macquarie), there are three different models available, and we take a weighted average of the three, with the weight determining the most appropriate one. This allows it to be used across all markets globally. As discussed previously, whilst Europe is very NAV-centric, the US market practice (due to using US GAAP rather than IFRS) is more focused on AFFO, whilst other REIT markets, particularly in Asia, are more dividend-orientated for valuations.

In this model, we take the current valuation for each of the three valuation methods, that is, earnings-based, dividend-based, and NAV-based. We then compare these valuations and fundamental growth rates with the peer group to determine an appropriate valuation for each of the three metrics. We then weigh each of the three target prices according to how relevant they are and use the weighted average as our target price.

The advantages of this method are 1) its simplicity and universal applicability, 2) it can be used by non-real estate specialists, and 3) the ease with which it can deal with both changes in share prices and changes in forecasts.

Figure 13.2 provides an example of how this would work in practice.

As can be seen, we have a share price of 100p, and historical and two-year forecasts for EPS(AFFO)/ DPS and NAV per share with subsequent valuations. We now go through the process of determining the appropriate multiplier of these per share figures.

13.2.1 *Earnings-based methods*

The current FYI Multiple is 14.29, that is, 100/7. How can we determine an appropriate multiple?

First, we could look at the long-term average multiple that the stock has traded at. Let us assume that this is 18. Therefore, if the REIT were to trade at its long-term average, the target price would be 18x the forecast earnings of 7p = 126p. In other words, if the valuation reverted to its long-term average and the forecasts remained the same, the shares would trade at 126p, a 26% uplift on the current share price.

Secondly, we could look at the current valuation relative to its peer group. A peer group would represent companies in a similar industry. To be truly comparable, they would also be of comparable size and have similar leverage. This is because, as Fama and French have shown, both size and leverage are key drivers of valuation. Let us say then that there are three similar companies, and their average multiple is 15. If our stock were to trade at a similar multiple, the forecast valuation and therefore target price would be 15 × 7 = 105p, that is, 5% above the current valuation.

Finally, we could look at the equity market overall. Let us say that the average multiple is 12. If our stock were to trade at an average equity multiple, the target price would be 12 × 7p = 84p, a 16% discount on the current share price.

We therefore have three different methods of determining an appropriate valuation. It comes down to our skill, expertise, and view of the market to determine which is the most appropriate one, not forgetting, of course, that it requires the rest of the equity market to agree with our view for the =target price to be achieved.

For the sake of argument, let us assume we go for option one. Therefore, our earnings-based target price is 126p.

13.2.2   Dividend-based methods

For dividend-based models, we could use the dividend discount model (DDM). This is not uncommon. However, a simpler method follows the approach outlined for earnings above. Therefore, we need to establish:

1 Long-term dividend yield average of the company – assume 4%
2 The current dividend yield valuations of the peer group – assume 3.5% for FY1.
3 Equity market and bond yield are comparable. Assume 3% for the equity market and 2.5% for 10-year bond yields.

The last step is to determine the most relevant comparable. Let us assume that we decide it is the peer group. We could argue that the company should trade closer to the same dividend yield as the peer group. If we chose a 4% target yield for FY1, this would still represent a premium to the peer group and be in line with the stocks' long-term average. To determine the target price, therefore, take the reciprocal of 4%, which is 1/0.04, that is, 25. We then multiply this figure by the FY1 dividend, which is 6p. This produces a dividend-based target price of 150p.

13.2.3   NAV-based methods

Because the NAV is specific to the listed real estate sector, we do not have the option of the third comparable of equity and bond markets. We therefore follow the first two steps.

The stock is currently trading at a discount of 23% to its FY1 NAV. Let us assume that its long-term average is 15%.

The peer group is trading at an average of 12%.

Therefore, we could assume that *ceteris paribus*, the discount could be expected to narrow from its current level. Let us assume it reverts to its long-term average of 15%.

To calculate the NAV-based target price, we would take a 15% discount (i.e., 1.00–0.15 = 0.85) and apply this to the FY1 forecast of 130p. This produces an NAV-based target price of 111p.

13.2.4   Combining the three methodologies

We now have the option of combining these three methodologies to produce a blended target price. We simply attach a weighting to each of the three methods as per Table 13.2.

Later in this chapter, we will discuss the sensitivity analysis which normally accompanies these methodologies.

13.3   Difference between sell-side and buy-side approaches

As mentioned, there are significant differences in the approach between the buy side and the sell side. For a sell-side analyst, the price target is explicit, will be shown on Bloomberg, and forms the logical underpinning of the investment recommendation. For the buy side, the price target is potentially less important and may well be replaced by a limit or review price which is a fixed % above and below the current price. The purpose of this is that when the stock reaches the upper or lower limit, the investment recommendation will be reviewed in light of the current market conditions. It is also important to note that whilst sell-side analysts deal in absolutes (Buy/Sell/Hold), the buy side is typically more concerned with weightings relative to a benchmark. They would not normally reduce the holding of a large stock completely, particularly in the case of a specialist real estate investor, where the weighting in the benchmark of that stock was material. Therefore, recommendations would normally be expressed as increasing or decreasing weighting.

13.4   Assumptions and catalysts that drive the target price

It is important to understand the diverse ways in which an analyst might want or need to adjust a target price before the anticipated date. Target prices are always refreshed when a company reports figures, but we are concerned here with changes outside the normal time for related reasons. There are two components to this.: the market valuation and the forecasted EPS, DPS, and NAV. It is the change in market valuation that normally has the most significant and immediate impact.

We can break these down into the following situations:

13.4.1   *A change in global capital market assumptions*

This would be an event that alters the forecasts for the economy and the risk appetite across asset classes. Examples would be the outbreak of war, other forms of geopolitical crisis, sharp increases in commodities – particularly oil–, unexpected Central Bank moves in setting interest rates, unexpected inflation, or GDP figures.

13.4.2   *A change in sentiment towards the sector as a whole*

Typically, generalist fund managers are more positive towards the sector and commit more capital to it at times of stable or reducing interest rates. As a result, a change in the expected trajectory of interest rates normally sees a sell-off and readjustment of equity market valuations in the sector.

13.4.3   *A change at the company level*

This could be because of a change of management or the market reaction to a new strategy, equity, or bond issue.

13.4.4   *A change at the asset level*

This affects the fundamentals and examples include disposals above or below market expectations, and a materially different outlook in a market update that necessitates a refreshing of the company forecasts by analysts.

13.5   Sensitivity of target prices and how to adjust them in response to market conditions

Given the volatility of equity market prices (+/−5% in one day for an individual stock whilst not common is not unheard of), it is sensible for the analyst to prepare a sensitivity analysis of what would happen to the target price if the value of any of the three different methods (EPS, DPS, NAV) were to change.

Figure 13.3 shows how this can be set up. A range of different multiples for each of the three methods is shown, and normally either the current price or the target price is in the middle of the table to illustrate the sensitivity. Looking at this table, for example, if we were to use a 100% weighting on the NAV method, then using a 40% discount would produce a target price of 78p and using a 5% discount would produce a target price of 124p.

EARNINGS	7	DIVIDEND	6	NAV	130
Multiple Range	Target Price	Yield Range	Target Price	Disc/Prem Range	Target Price
11	77	4	150	-40	78
12	84	4.5	133	-35	85
13	91	5	120	-30	91
14	98	5.5	109	-25	98
15	105	6	100	-20	104
16	112	6.5	92	-15	111
17	119	7	86	-10	117
18	126	7.5	80	-5	124

Figure 13.3 Sensitivity analysis of target price.

Source: The authors.

13.6   How to make a logical investment recommendation

The final step is to make an investment recommendation. At its simplest, this could be based on a simple difference between the calculated target price and the current share price. If the target price is >10% above the current share price, then the recommendation is a buy. If the target price is <10% below the current share price, the recommendation is a sell. This produces a recommendation based on absolute figures. It is also possible to produce a recommendation based on expected relative performance to a benchmark. It is worth noting that the methodology for recommendations will vary from investment bank to investment bank and by geography. As an example, this is the range of recommendation categories used at one investment bank.

Australia/New Zealand
Outperform – return >5% more than benchmark return.
Neutral – return within 5% of benchmark return.
Underperform – return >5% below benchmark return.

Asia/Europe

Outperform – expected return >+10%

Neutral – expected return from −10% to +10%

Underperform – expected return <−10%

South Africa

Outperform – expected return >+10%

Neutral – expected return from −10% to +10%

Underperform – expected return <−10%

Canada

Outperform – return >5% more than benchmark return.

Neutral – return within 5% of benchmark return.

Underperform – return >5% below benchmark return.

USA

Outperform (Buy) – return >5% more than Russell.

3000 index return

Neutral (Hold) – return within 5% of Russell 3000 index.

return

Underperform (Sell)– return <5% below Russell 3000 index.

return

Quiz questions

1 What is the purpose of a target price?

2 Name two methods for determining a target price which do not involve using a weighted average?

3 What are the three methods of valuation used in a weighted average target price methodology?

4 What are the four main reasons an analyst might change a target price before the anticipated date?

5 How can you determine an investment recommendation based on a target price?

Discussion questions

• Do you think sell-side target prices are meaningful?

• How do you determine the most relevant methodology for a target price?

Research/Dissertation topics

• How would you measure the accuracy of target prices and recommendations?

14 Operational mechanics of a public real estate company

Objectives

At the end of this chapter, you will understand:

- How a listed real estate company operates.
- Corporate strategies for growth.
- The natural evolution of a company's shareholder base.
- What happens on the day a company's financial results are released.

Key concepts

- The different functions and responsibilities within a listed real estate company.
- Different corporate strategies available and their impact on real estate decisions.
- How a company's shareholder base evolves as it grows.
- Distinction between regulatory and discretionary company announcements.

14.1   How a listed real estate company operates

There are four main internal operations of a listed real estate company:

- **Asset management**
  Optimising the value of the real estate portfolio
- **Liability management**
  Ensuring a prudent and optimal financing structure is in place to enable the real estate strategy to be implemented.
- **Corporate management**
  Whether internally or externally managed, there is an obligation on the management team to ensure that the corporate level is administered efficiently without excessive overheads.
- **Investor relations/stock market regulations**
  Maintaining relations with investors/analysts and ensuring compliance with stock market regulations in terms of disclosure of information.
  Interestingly, many companies have set up a fifth operation – ESG – but in most cases, whilst this may have a separate team, it influences all four elements above and its influence is increasingly holistically integrated within the company.
  We now turn to external elements and relationships. For our purposes, we can separate these into the following categories:

DOI: 10.1201/9781003298564-17

- **Retained financial adviser: corporate broker/investment bank**

  In Europe, every listed company has to have one or more firms named as its financial adviser. This is not the case in the US, where companies can use different advisers from transaction to transaction.

  In the UK, there is a distinction for main market listings between the stockbroker and the financial adviser. For the sell side, building up a large roster of successful companies as corporate clients is seen as key to ensuring the consistency of new business (equity issues) and cementing relationships with shareholders due to the frequency of both primary (IPOs) and secondary (aka seasoned equity offerings) issues. When a company decides to float/list/ IPO, it will typically have a beauty parade of firms invited to pitch for the business. Smaller companies tend to pick one domestic firm (at least initially), whereas, for larger issues, it is often the case that there is a multitude of firms involved – known as a syndicate – at various levels, from global coordinator (the highest) to a country or regional book-runner. In case you are wondering whether so many sell-side firms need to be involved, the answer is that typically – apart from exceptionally large issues, – they do not, as there will be enormous overlap of buy-side clients between them. The reason a company appoints so many is that any investment analyst who works for a firm which is part of the syndicate is not deemed to be independent and therefore must have research approved by the syndicate. Therefore, the greater the number of firms involved, the fewer independent analysts there are to write potentially negative comments about the company ahead of the transaction.

  The company will be in very regular contact with its appointed advisers once it is listed, and the buy side will typically put a high weighting on the company broker's analyst forecasts, as they will be seen as having the greatest access to the management team and have had the forecasts carefully reviewed by the company. This "home team advantage" is seen as a boost to the analyst's credibility.

- **Non-retained financial product providers: investment banks**

  Once a company has appointed a financial adviser and is listed, it is still free to use the services of other firms. This often occurs if investment banks are marketing a specialist product, or the company requires access to a specific pool of investors. For example, if Goldman Sachs developed a particularly innovative and subsequently popular form of convertible bond, a company might use them specifically to raise money via this instrument, as they would have experience of the investors investing in this instrument, which is neither pure equity nor pure debt.

- **Retained PR team**

  In addition to the retained financial adviser, there will normally be a public relations firm, which advises the company on press releases, discretionary announcements, investor presentations and market feedback. They will also coordinate meetings with journalists.

- **Retained lawyers**

  Most firms have either one or a group of lawyers which they normally use. These are typically selected based on their specialist knowledge in certain areas and can be either formally or informally retained. Larger companies will have a house legal team.

- **Financial adviser investment analyst**

  As mentioned, the investment analyst at the firm which acts as the financial adviser to the company will have greater access to and dialogue with the firm than independent analysts. As such, the research output of the firm will also carry out a disclosure that the firm acts as a financial adviser and therefore is not independent. The buy side will typically look at the "house broker" as a starting point for forecasts for the company, as they will be deemed to have been subject to additional scrutiny. As they are not independent, the recommendation will always be assumed (and therefore discounted) to be positive. And, indeed, it is almost unheard of for a house broker to issue a sell recommendation on a client company.

- **Independent investment analysts**

Subject to the size of the company, there may be between 1 and 30 independent analysts who will decide to produce recommendations and forecasts on a company. The key factors will be the expected level of turnover in the shares and the possibility of gaining corporate finance business from the company.

- **Paid-for investment analysts (smaller and newer companies only)**

Post Mifid 2, a recent development has been the advent of analysts producing research on a company, which is paid for by the company. There are two reasons for this: first, fewer analysts write on small companies as commissions are lower due to their size and the (in)frequency of trading. Second, smaller companies often seek to target individual share-holders rather than the institutions, which are the clients of the larger broking firms. Therefore, paid-for research has become commonplace. It is like a house broker's note but without a recommendation and can be distributed to anyone, as it is not a solicitation to buy shares, merely information provided on the company.

- **Existing shareholders**

Once a company is listed, its priority, when its results have been released, will be to visit its existing shareholders in a series of one-to-one meetings known as a roadshow. For larger companies, this can take up to three weeks following the results and could include North America, Europe, and Asia. More typically, the roadshows are limited to one region. Whilst no new information can be given at these meetings (that would be "inside information"), the purpose is to allow the company to provide some back-ground or "colour" on the figures and for the buy side to make a subjective assessment of the management credibility of the team. This is normally organised by the company's retained adviser/broker.

- **Potential new shareholders**

In addition to existing shareholders, a company will also visit potential new shareholders to convince them of the investment case for buying their shares. Often the institution will not buy straightaway but will commit over the medium term or on placing a corporate transaction.

- **Journalists**

Finally, there are the financial journalists. The PR firm will normally organise a press conference (sometimes combined with an analyst's conference) on the day of the figures. This may be supplemented with longer interviews with journalists during the year.

14.2   Strategies for growth

A company's specific corporate strategy will be dependent, *inter alia*, on its existing size, the preferred sector, competition, and corporate structure. For example, the strategy of Prologis, the largest logistics provider in the world, based in the US, will be different from Warehouse REIT, a small UK company, even though they are in the same sector.

In this section, we look at some examples of different corporate strategies, using some case studies to illustrate the assorted styles available. We look at their style (how they are perceived by investors), their aim (their story to investors) and their valuation objective, which allows them to achieve this strategy. We have chosen five specific styles, but there are many more available, and in certain countries, they may be expressed differently even if the objectives may be similar. The corporate strategy can evolve (see next section) and will be at least partly driven by asset and company size, maturity/liquidity of the assets, quantum and quality of legacy assets, issues in the debt/liability portfolio, vehicle structure (REIT/Non-REIT) and the management team.

The five chosen strategies are:

- Specialist aggregator
- Specialist major
- Investment stylists
- Beta plays
- Adaptors

Specialist aggregator

- **Style:** A niche player entering with a view to achieving critical mass.
- **Aim:** Convince investors of the merits of the asset class and the management team.
- **Valuation objective:** Trade at a premium to issue more equity.
- **Examples:** Tritax Big Box (historically), Warehouse REIT.

Specialist major

- **Style:** A niche that has achieved critical mass
- **Aim:** Convince investors it can continue to grow and be the largest and highest quality play in the sector
- **Valuation objective:** Trade at a premium to peer group
- **Examples:** Unibail (historically), Derwent London

Investment stylists

- **Style:** A portfolio designed *not* for sector exposure but to provide an investment objective, for example, income, and inflation protection.
- **Aim:** Convince investors it can continue to meet their specific requirements via real estate.
- **Valuation objective:** Provide sufficient income/dividend growth, etc., relative to other equity sectors.
- **Examples:** Secure Income, Supermarket REIT.

Beta plays

- **Style:** A portfolio designed for exposure to the diversified real estate sector in general
- **Aim:** Convince investors it can produce real estate returns with liquidity but without too much equity noise
- **Valuation objective:** To beat MSCI/IPD indices and produce a predictable total return.
- **Example:** AEW REIT

Adaptors

- **Style:** A portfolio originally based in one sector (offices or retail) but gradually repositioned towards more growth areas
- **Aim:** Convince investors it has specialist skills in the new areas
- **Valuation objective:** To beat returns from previous single-sector peers.
- **Examples:** Cofinimmo, Wereldhave, Gecina

14.3   Evolution of a shareholder base

As a company grows in terms of market capitalisation the shareholding base will change/evolve to larger institutions that can support further equity raises. In addition, the company will enter more equity indices and the number of generalists on the register will increase, as will the liquidity of the shares.

Broadly speaking we can divide this evolution into five stages. Below we show these five stages in terms of shareholder types and valuation drivers to illustrate the change as the company grows.

1 **Pre-IPO**
   Shareholder types
   This could take the form of founding directors, friends and family, or a private equity house.
   Valuation drivers
   The exact mix will have an impact on IPO pricing and fundraising, as the key metric will be new money raised.
2 **IPO to junior market**
   Shareholder types
   Key investors are (subject to the size of the IPO) likely to be individuals, wealth managers, specialist funds or generalist-focused funds, for example, income funds.
   Valuation drivers
   A discount to existing comparable stocks to allow for a small premium on listing.
3 **Move to main market and EPRA indices.**
   Shareholder types
   Increasing numbers of specialist real estate investors and, for the first time, subject to size and liquidity, index-based active and passive funds.
   Valuation drivers
   Assuming a track record, valuation will start to become company-specific and closely linked to the underlying asset type.
4 **Inclusion in general equity indices**
   Shareholder types
   As the company grows and is included in more equity indices, the level of generalist investors and sell-side analysts following the company is expected to increase.
   Valuation drivers
   Still closely related to property fundamentals, but increasingly the role within the peer group is to finetune valuation targets from analysts and specialist investors.

5 **Leading company in sector/market**
   Shareholder types
   Once the company reaches critical mass, there will be a limit to potential shareholders, including short sellers (subject to minimum market cap)
   Valuation drivers
   Increasingly influenced by market sentiment and fund flows

In this section, we have:

- Identified the various stages of a company's growth.
- Understood the key investor targets.
- Demonstrated how the valuation drivers for a company change during its evolution.

14.4   The day company figures are released ("A Day in the Life")

In this section, we look at exactly what happens in terms of disclosure and assimilation of information on the day a company announces its financial statements (we assume full-year results as they are the most important.). It is important to understand that all relevant and sensitive information must be released by a company to the market as a whole and as soon as possible.

Examples of this disclosure include transcripts and videos of results presentations, announcements ahead of group meetings, for example, capital markets day and quarterly updates on trading.

We now look at the timeline following the release of the company's figures via the stock exchange, which occurs typically at 7 am.

We have divided the periods into the time following the announcement and show the questions that both buy and sell-side analysts will be asking. Remember sell-side analysts will have a detailed financial model and will be feeding the new figures and their assumptions into this model to produce revised forecasts and target prices/recommendations.

First 5 minutes

- What is new or unexpected in the statement?
- Is this positive or negative for the share price?
- What is not in the price?

First 30 minutes

- Examining the tone of the statement
- What has not been said and quantifying the potential impact?

Rest of the day

- Digesting the company presentation
- New forecasts and price targets after given new information and share price action.

Once the market has opened

- Analysts' meeting where the management team go through the figures in more detail, have a presentation to accompany the results, and answer questions. This meeting is normally recorded, and a transcript made available via the stock exchange news service.
- Shareholder meetings are often split by region, for example, UK, European and US. These are targeted at both existing and potential shareholders.

Key questions that will be sought by investors and analysts at the meetings.

- Does the management team have credibility?
- Does the management team seem to be in agreement?
- Are they listening to the questions?
- How do they respond to criticism/suggestions?
- Can they make the right decisions in difficult times (GFC, EU referendum, COVID, etc.)?
- Are they problem solvers/value creators/alpha generators or administrators?
- All companies are now asked specifically about their ESG credentials and strategy. Most institutions want to know how the process is implemented from the top down, as well as specific details on carbon emissions reduction.
- This subjective analysis forms a key part of the valuation process and directly impacts a company's fundraising ability.

From the company's viewpoint

- The internal IR team and their corporate advisers have a significant role in managing the information flow and ensuring that their key messages are understood.
- The best way for companies to respond is by incorporating analysts' suggestions and sensitivity analysis tables in their releases to address issues.

14.5   Understanding what your share price is telling you

Finally, we look at some typical questions asked by companies and provide answers using the material covered in this book.

"How come our share price has gone down when our results were well up on last year?"
Share prices have built on expectations. When a company releases its results, it means that there is an objective way of telling whether those expectations have been met or not. It also means that there will be no new information for a period. It is therefore common for investors to buy shares ahead of figures and sell on the announcement.

"Why have Blackrock bought and sold our shares on the same day?"
The same fund manager (in this case Blackrock is used as an example) can have many different types of funds under their name. It is therefore not uncommon for say an income fund to sell shares in a company whilst a value fund in the same fund management company might buy them. As far as the funds are concerned, they are independent entities. Therefore, careful analysis of the shareholder register needs to be undertaken to determine which funds are included in the same group.

"XYZ is trading at a 65% discount to NAV: that must be cheap, and they must be a BUY."
A "high" valuation (e.g., premium to NAV) does not necessarily mean that a REIT is "expensive", merely that growth expectations are high. Similarly, a stock at a big discount is not necessarily "cheap" and it is the Forecast Year 2 NAV that is important, not the historical one.

"Following the full-year figures, ABC analyst changed his price target to 270p and his recommendation from sell to buy. XYZ Analyst changes his price target to 230p and reiterated his sell recommendation." Who is right?
There are many different methods of calculating a target price and they are not directly comparable.

"I don't understand it; we had a solid shareholder base, but they didn't support our third equity raise this year."
A company needs to manage its shareholder base so that it has the correct balance between liquidity and institutions with the ability to support its equity capital-raising ambitions.

Quiz questions

1 What are the four main operational elements for a real estate company?
2 Name three types of investment analyst that would follow a listed real estate company.
3 List five different investment strategies for a listed real estate company.
4 What are the five stages of a listed real estate company's evolution?
5 What two factors would you use to determine the stage of a company's evolution?

Discussion questions

- Identify different companies within the same sector (e.g., logistics, offices) and determine their current and optimal corporate strategy.
- How would you change the approach of your investor presentation when dealing with a generalist rather than a specialist investor?
- Look at a recent corporate transaction and determine the roles of the different investment banks involved in the transaction.

Research/Dissertation topics

- Liability management is as important as asset management in determining long-term corporate valuation and fundraising ability.

Applications

15 Public market applications

Objectives

At the end of this chapter, you will understand:

- The structure and purpose of dedicated Public Real Estate Securities Funds.
- Combining listed real estate with other assets.
- The blended approach to real estate allocations.

Key concepts

- How the risk and return of public real estate allocations impact portfolio returns.
- The time-variant nature of correlations.
- Geographical differences in application.
- Differentiation of investment strategies and investment processes.

15.1 Structure and purpose of dedicated real estate securities funds

When we talk about dedicated real estate securities funds, we are referring to equity funds that only invest in public real estate. If, for example, they are combined with other assets such as commodities, infrastructure equities and possibly inflation-linked assets, they would be classified as a real assets fund, not a real estate securities fund. Within this broad classification of dedicated real estate securities funds, there are several different investment strategies and structures available. We will deal with the geographic breakdown shortly. The key asset management strategies available are (Figure 15.1):

Long-only or long-short
Actively or passively managed.
Benchmarked or non-benchmarked
Use of relative or absolute benchmark

Although listed real estate has been a feature of capital markets (in the US, Europe, and Australia) since the 1960s, the first example of a dedicated global real estate securities fund that we are aware of is what is now known as the Alpine International Real Estate Fund, launched in 1989 by Sam Lieber. Given the growth in the US REIT market, which started in the early 1990s, this proved to be a prescient move.

Starting in the late 1990s and growing most notably in the period 2000 to 2007, there has been a dramatic expansion of dedicated real estate securities funds (both regional and

DOI: 10.1201/9781003298564-19

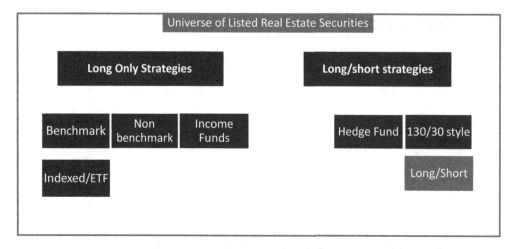

Figure 15.1 Variety of investment strategies for dedicated real estate securities funds.

Source: Moss and Baum 2013, The use of listed real estate securities in asset management.

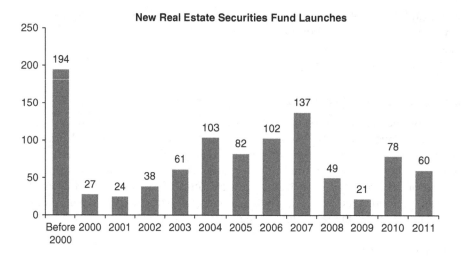

Figure 15.2 Growth of real estate securities funds.

Source: Consilia Capital.

global), as well as unlisted funds. Figure 15.2 shows the number of new real estate securities funds launched annually between 2000 and 2011.

The roots of the growth in real estate funds (listed and unlisted) in the period 2000–2007 lay in the superior performance of the underlying real estate market during that period, which, fuelled by strong occupational demand and an increase in available debt capital, saw the asset class produce a comparable return to equities with lower volatility. This superior risk-adjusted return resulted in both retail and institutional investors seeking exposure to the underlying asset class. In addition, the globalisation and securitisation that was occurring, led by the introduction of REIT legislation across the world, increased the geographic scope, absolute size, and liquidity of the investable universe. Product developers employed by asset managers were able to create collective investment vehicles that allowed this pool of capital to be harnessed in a variety of different ways.

Although there are clear differences between the structure of an unlisted fund (be they core, core-plus, value-added or opportunity) and an index benchmarked listed fund, the investment rationale is often similar. In summary, there was significant demand from both retail and institutional investors to gain access to the underlying asset class via a collective investment vehicle.

Although the number of new funds being launched has now slowed down and there is increasing pressure on fund management margins in this area, it is interesting to note that after 2007, there have been some new entrants to the sector. At the larger end of the spectrum, for example, Blackrock has announced the creation of a global real estate securities platform and the formation of regional teams. In the UK, listed real estate funds and capabilities have recently been set up at asset managers with direct property experience, most notably Grosvenor Fund Management, Tristan Capital and Internos.

However, in common with other asset classes, institutional and retail investors are re-assessing appropriate asset class weightings since the global financial crisis. Given the low level of risk-free asset rates globally, current low levels of inflation and the need to maintain flexibility in portfolio allocation, there is a growing trend to focus on the underlying investment characteristics (such as certainty and level of income return and liquidity) of the asset/investment product rather than obtaining blanket exposure to an asset class. In addition, there is increasing interest in how combining different asset classes may produce required risk-adjusted returns.

It is against this background of new risk/return requirements and investment vehicles that the liquid, listed real estate sector can contribute to portfolios. Our starting point is an examination of the range of combinations available to portfolio managers and product developers using listed real estate as all or part of the real estate portfolio. We can think of this as a palette of available real estate options with subtly different risk and return characteristics compared to a benchmarked listed real estate exposure. Having identified the available options, we look at each in turn to see what investment products are currently available, how they differ, and which areas are likely to increase because of regulatory or market factors.

15.1.1 *Which benchmark do these funds use?*

When seeking to determine the different strategies that funds operate, the easiest starting point is to see whether a benchmark is used. If so, investors have a clear indication of the desired risk/return profile and investment objectives of the fund. For the funds that used a benchmark, EPRA was by far the most popular provider, accounting for two-thirds of funds using benchmarks.

Funds have different approaches on how far their weightings will differ from the chosen benchmark, which is dependent on several factors, including the following:

  i the level of tracking error that investors are happy to accept, as indicated in the fund prospectus.

ii the size of the fund: the smaller the fund, the less likely it will be that realising individual positions will have an impact.

iii whether the portfolio is constructed using the benchmark regional weightings as a starting point or whether the portfolio is assembled "bottom-up" by selecting the most attractive company valuations regardless of region.

iv the frequency with which the fund trades.

v whether the fund is open-ended and subject to inflows and redemptions from unit holders which affect portfolio weightings and activity, or close-ended.

vi whether the fund has a secondary objective such as a preferred minimum yield, liquidity, or currency requirement.

vii the number and location of staff involved in the fund. It would be unusual, for example, to take a significant underweight position in a region where most staff were located.

viii the time period.

For the purposes of performance attribution, there are several factors which can lead to under/outperformance, and these factors will vary over different stages of the cycle as well as being dependent on the reporting currency of the fund.

In addition to benchmarked funds, there is also a category of non-benchmark-constrained funds. These funds seek to deliver returns related to the underlying real estate of their holdings but are not concerned with tracking errors relative to a benchmark.

15.1.2 *What is the geographic breakdown of these funds?*

We are now aware of the different styles of funds, so we can examine their geographic exposure. In broad terms, real estate securities funds can be split into Asian, European, US and global mandates.

Figure 15.3 shows the breakdown between each of these mandates at the end of 2019 in terms of AuM, number of funds, volatility, returns and tracking error. We have also shown funds that are deemed by asset allocators to be similar, namely real estate income funds, global and regional infrastructure funds, and alternative strategies.

Of these, the most interesting are the alternative strategies. Figure 15.4 shows how these were broken down at the end of 2019.

The difference between these and traditional funds based on a free float market capitalisation index benchmark can be seen as:

Smart beta strategies: Equal weighting
Sector focus: Housebuilders, residential
Investment themes: Cities, sustainable, leaders, small cap
Enhanced passive strategies: leveraged bear and bull, inverse.

15.2 Combining public real estate with other assets

Having examined how public real estate can be used as a dedicated asset class, we now turn to how it can be combined with other assets, as shown in Figure 15.3 (Figure 15.5).

15.2.1 *Listed real estate fund with direct property*

The most well-known example of combining listed and direct property in a fund format is the *TR Property Investment Trust*.

Mandate	Assets under Management	Number of Funds	Volatility (%)	2019 return (USD%)	5 yr return (USD%)	Tracking Error (%)
Asian Real estate	2,318	42	12.9	18.6	37.5	5.8
European real estate	17,759	66	13.7	25.4	39.6	3.9
US Real estate	162,637	102	14.4	24.1	31.0	5.8
Global Real Estate	80,172	220	10.6	22.5	26.8	4.0
Real Estate TOTAL/AVERAGE	**262,886**	**430**	**12.9**	**22.7**	**33.7**	**4.9**
Income	23,811	35	10.7	25.7	37.2	4.9
Global Infrastructure Fund	44,347	98	10.3	25.1	31.4	4.7
Regional Infrastructure	4,104	36	14.3	11.2	24.0	7.8
Alternative Strategy	25,579	70	11.7	19.6	21.3	8.9
Other TOTAL/AVERAGE	**97,841**	**239**	**11.8**	**20.4**	**28.5**	**6.6**
Grand Total	**360,727**	**669**	**11.9**	**22.4**	**30.0**	**5.4**

Figure 15.3 Real estate securities universe.

Source: Consilia Capital.

Data

Size	Assets under Management	Number of Funds	Volatility (%)	2019 return (USD%)	5 yr return (USD%)	Tracking Error (%)
Active ETF	119	1	13.96	28.77	45.16	1.18
Cities	1,256	1	10.52	30.26	30.90	5.88
Equal Weights	343	2	15.96	24.33	24.38	0.88
Housebuilders	775	1	16.66	41.30	38.83	0.45
Inverse	7	1	12.81	-20.96	-35.11	22.26
Leaders	2,337	1	13.78	25.46	42.36	0.33
Leveraged Bear	33	2	33.24	-45.34	-71.22	39.96
Leveraged Bull	196	3	31.20	62.14	67.00	15.72
Leveraged Mortgage REIT	562	2	21.48	37.62	91.77	10.10
Mortgage REIT	1,429	1	11.49	21.38	54.49	0.51
Multifactor	8	3	9.85	27.43	48.70	1.06
Real Assets	17,202	46	8.81	17.73	17.37	8.36
Real Assets eTF	11	1	9.19	24.41	n/a	1.91
Residential	508	1	14.73	24.47	54.47	0.65
Small cap	65	1	14.09	24.14	24.82	0.74
Sustainable	72	2	8.85	16.05	15.86	3.21
TOTAL	**25,579**	**70**	**11.75**	**19.59**	**21.26**	**8.95**

Figure 15.4 Alternative strategies.

Source: Consilia Capital.

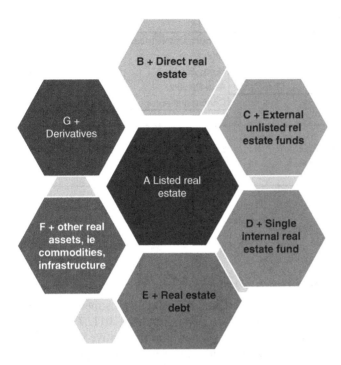

Figure 15.5 Palette of options available for combining listed real estate with other assets.
Source: Moss and Baum, 2013: The use of listed real estate securities in asset management.

The investment guidelines are as follows:

"To deliver a spread of investment risk, the maximum holding in the stock of any one issuer or of a single asset is limited to 20% of the portfolio. In addition, any holdings more than 5% of the portfolio must not in aggregate exceed 50% of the portfolio. These limits are set at the point of acquisition. However, if they were materially exceeded for a significant length of time through market movements, the manager would seek to remedy the position."

The manager currently applies the following guidelines for asset allocation:

UK-listed equities 25–50%
Other listed equities 0–5%
European listed equities 45–75%
Listed bonds 0–5%
Direct UK property 5–20%

15.2.2   *Listed real estate fund with unlisted funds*

A good example of a structure combining listed funds with unlisted funds is the *HSBC Open Global Property Fund* run by Guy Morrell. This combines a geographic allocation strategy with the ability to invest in funds that are exposed to both listed and unlisted property markets. Figure 15.6 shows the allocation of this fund as of October 2012.

HSBC Open Global Property Fund Allocations October 2012

Asset Allocation	%	Geographic Allocation	%
Property Securities Funds	61	UK	42
Direct Property Funds	37	North America	27
Cash	2	Asia Pacific	26
		Continental europe	5

Top 5 Holdings	%		%
HSBC MultiAlpha Global real Estate Equity	11	Schroder Global Property Securities	10
Schroder Asia Pacific Property Securities	11	Henderson UK Property Class	9
		M&G Property Portfolio	9

Figure 15.6 HSBS fund allocations.

Source: HSBC Open Property Fund allocations.

HSBC Fact Sheet.

15.2.3   *The Legal and General Hybrid Property Fund*

Legal & General Property (LGP) together with Legal & General Investment Management (LGIM), launched a new product, the Hybrid Property Fund, in 2011. They claim this offers defined contribution (DC) pension schemes a new and innovative way to invest in property funds while managing volatility and liquidity. Developed in conjunction with an investment consultant, the fund invests in LGP's UK balanced fund (the Managed Property Fund) and LGIM's Global REITS Index Tracker Fund on a default position split of 70:30. As of May 2011, the fund was available to several other DC schemes and has been selected by a major life platform to provide its clients with global real estate exposure.

Providing a property allocation that has been specifically designed in conjunction with investment consultants to meet the optimum criteria of UK DC pension schemes, L&G says the fund caters to the increasing long-term growth trend of DC. The combination of UK direct and global indirect funds provides greater diversification and liquidity while reducing fund expenses and the entry and exit costs typically associated with direct property investment. To provide the ability to adapt to market movements, the manager has the flexibility to alter the 70:30 default position of the fund within pre-set benchmark ranges.

15.2.4   *F: Listed real estate with other "real assets"*

The growth in commodities and resources as an asset class, coupled with a need to provide protection against inflation, has led to increasing attention on "real asset" funds.

15.2.4.1   *The Cohen & Steers Real Assets Fund*

Cohen & Steers hold the view that massive global stimulus and sustained growth in emerging markets will ultimately drive inflation higher. They believe real assets can offer attractive investment characteristics from a fundamental perspective, along with inflation-fighting tendencies that help maintain a portfolio's long-term purchasing power.

The structure of this fund, which we believe may be replicated by others, is as follows: It invests primarily in core real asset categories, consisting of global real estate securities

(25–35%), commodities (25–35%) and global natural resource equities (15–25%). It may invest up to 20% of the portfolio in diversifiers for added stability, including gold and fixed income in multiple currencies. It utilises a multi-manager approach for core real asset categories.

15.3 Blending public and private real estate

One of the key challenges for both asset allocators and product developers is how to provide real estate exposure in a mixed asset portfolio with acceptably high levels of liquidity and low levels of cost. A 100% exposure to unlisted funds or direct real estate would not be expected to meet this demanding criterion. Therefore, a blended approach is used, that is, combining direct real estate exposure with public real estate. The key questions that would determine the efficacy of this approach would be as follows:

Return enhancement: What is the "raw" performance impact of adding listed real estate to an unlisted portfolio?

Risk-adjusted impact: What is the impact on portfolio volatility and the Sharpe ratio?

Tracking error: Does adding a global listed element significantly increase the tracking error of the portfolio relative to a UK direct property benchmark?

Currency impact: Does adding a global listed portfolio introduce a material currency risk into portfolio returns?

Cash drag: What is the impact on returns and volatility of adding cash to the portfolio?

Risk attribution: What adjustments are necessary to understand the true relative contributions to portfolio risk?

Portfolio contribution: Does this blended real estate product provide the diversification benefits of real estate in a multi-asset portfolio?

In 2014, Moss and Farrelly undertook a study to answer these questions. They looked at combining a direct UK real estate exposure (70% of the portfolio) with a 30% exposure to a global REIT tracker.

The results are shown in Figure 15.4 and are broken down into different stages of the cycle (Figure 15.7).

There was a positive return enhancement overall, even though the majority of these occurred in one period, the QE-led recovery.

		Total Returns			
Period	Dates	UK Unlisted Funds	Global Listed Funds	70:30	Return Enhancement From Adding Listed
TMT Boom & Crash	June 1998 - June 2003	65.5	33.9	56.0	-14.4%
Rising UK Property Values	July 2003 - June 2007	81.7	107.7	88.4	8.2%
Global Financial Crisis	July 2007 - June 2009	-33.0	-34.5	-33.5	-1.3%
QE Led Recovery	July 2009 - June 2013	32.3	103.6	52.2	61.6%
Past Five Years	July 2008 - June 2013	4.1	62.6	20.3	390.6%
Past Ten Years	July 2003 - June 2013	59.7	154.8	85.6	43.3%
Full Period	**June 1998 - June 2013**	**166.4**	**270.8**	**197.7**	**18.8%**

Figure 15.7 Returns over the cycle.

Source: Moss and Farrelly (2014).

	UK Unlisted Funds	UK Unlisted Funds Inc Subscription Costs	70:30 UK Unlisted Funds: Global Listed Funds	70:25:05 UK Unlisted Funds: Global Listed Funds:Cash
Portfolio Allocation				
Unlisted Property Funds	100%	100%	70%	70%
Global Listed Funds	0%	0%	30%	25%
Cash	0%	0%	0%	5%
Portfolio Statistics				
Annualised Mean	6.8%	6.4%	7.7%	7.1%
Annualised Geometric Mean	6.8%	6.3%	7.5%	7.0%
Annualised Volatility	6.4%	6.5%	8.4%	8.0%
Beta vs IPD Monthly Index	0.88	0.88	0.93	0.88
Tracking Error vs IPD Monthly Index	1.3%	2.0%	5.4%	5.2%
RSq with IPD Monthly Index	0.97	0.92	0.60	0.60
Sharpe Ratio	0.67	0.60	0.62	0.58
Modified Sharpe Ratio	0.35	0.32	0.33	0.19
Information Ratio - IPD Monthly Index	-0.34	-0.42	0.08	-0.02

Figure 15.8 Comparison of results for blended portfolios.

Source: Moss and Farrelly (2014).

Although return enhancement is clear, it is important to understand the additional risk and the tracking error that was introduced by adding public to private real estate.

Figure 15.8 shows the breakdown of the key risk and return metrics over the full period for the various portfolio combinations examined. Starting with a pure unlisted fund, subscription costs are added, then a 30% public element, and finally a 25% public 5% cash element.

The results and therefore answers to the seven key questions are as follows:

Return enhancement: Over the past 15 years, a 30% listed real estate allocation has provided a total return enhancement of 19% (c. 1% p.a. annualised) to our unlisted real estate portfolios. Over the past ten years, this was 43% (c. 2% p.a. annualised), a result which is consistent with the previous Consilia Capital study. Over five years the enhancement is c. 4% p.a. annualised, amounting to +390% in absolute terms.

Risk-adjusted impact: The price of this enhanced performance and improved liquidity profile is, unsurprisingly, higher portfolio volatility of around 2% p.a., from 6.4% to 8.4%. However, because of the improved returns, the impact on the Sharpe ratio is limited.

Tracking Error: We found that there is an additional 4% tracking error cost versus the direct UK real estate market when including 30% listed allocations. We believe that this is surprisingly small given that the listed element comprises global rather than purely UK stocks. We also found that a c. 1.3% tracking error arises for a well-diversified unlisted portfolio, highlighting that a pure IPD index performance is unachievable. This tracking error rises to 2% if subscription costs are included.

Currency impact: We found that the annual difference in returns and volatility between a hedged and an unhedged global listed portfolio over the 15-year period of the study was not material.

Cash drag: We found that the impact of adding a 5% cash buffer to the portfolio was to reduce annualised returns over the period by 0.6%, from 7.7% p.a. to 7.1%, and to reduce volatility from 8.4% to 8%.

Risk attribution: While the volatility of listed exposure is well-known, it is equally well recognised that the true volatility of unlisted funds is greater than commonly stated. We refined our measurements for risk by accounting for non-normalities and valuation smoothing and found that unlisted funds contributed to a greater share of overall risk.

Portfolio contribution. We modelled the impact of using our DC Real Estate Fund rather than a 100% unlisted exposure in a mixed asset portfolio of equities and bonds. The impact was extremely similar and marginally better if unsmoothed data was used as a comparable, modestly raising the Sharpe ratio for the mixed asset portfolio over the 15-year period, whether a 10% or a 20% real estate weighting was used.

This study therefore demonstrates the benefit of incorporating a listed real estate element into the real estate allocation. In addition to enhancing performance, this satisfies the market product requirements for liquidity and low cost. Risk decomposition analysis highlighted that if downside risks are considered, the incremental risk of including listed real estate is overstated when focussing solely on volatility. DC Real Estate solutions would still provide good diversification benefits to multi-asset portfolios.

The next steps in the evolution of this integrated approach will be related to the public element of this hybrid approach. We expect them to revolve around the use of smart beta strategies, active tilts, and mechanical trading rules, rather than a simple buy-and-hold strategy, be it active or passively managed.

Quiz questions

1 List four ways in which the structure of a dedicated real estate securities fund might differ.
2 What accounted for the growth in real estate securities funds from 2000?
3 What factors determine how much a fund manager might allocate weightings away from the benchmark?
4 List six other asset classes that public real estate might be combined with.
5 What are the seven ways in which you would determine whether adding public real estate to a private allocation had been successful?

Discussion questions

• At what stages of the cycle does it make sense to increase the weighting of public real estate securities in a blended allocation?
• Can the same team manage public and private real estate allocations?
• How would you seek to outperform an EPRA benchmark?

Research/Dissertation topics

• What smart beta strategies could be employed to outperform a free-float market-weighted benchmark index?

Appendix

Appendix
Quiz questions and answers

Chapter 2 Understanding different shareholders' objectives

1 **Name five of the largest real estate managers globally.**

Blackstone
Brookfield Asset Management
MetLife Investment Management
PGIM Real Estate
Nuveen
ESR Group
JP Morgan Asset Management
CBRE Investment Management
AXA IM Alts
Starwood Capital Group

2 **What is a Fund of Funds/Multi-Manager?**
These are groups who provide advisory services (discretionary and non-discretionary) to institutional clients and allocate capital to different (predominantly) externally managed funds. They play an important part in the capital allocation process.

3 **What are the two types of listed real estate manager?**
The first type is the dedicated REIT portfolio managers such as Cohen & Steers. They manage funds which hold listed real estate. The second type includes names such as CapitaLand. They own asset management platforms which manage publicly listed REITs, and they may also take direct stakes in listed real estate companies.

4 **What are the six different classifications of investors?**

1 Private individuals
2 Wealth managers
3 UHNW wealth managers
4 Institutional asset managers
5 Pension funds
6 Sovereign wealth funds

5 **Which countries have the largest sovereign wealth funds?**

China Investment Corp (CIC)	China
Norges Bank Investment	Norway
Abu Dhabi Investment Authority (ADIA)	UAE
State Admin of Foreign Exchanges	China
Kuwait Investment Authority	Kuwait
Got of Singapore Investment Corporation (GIC)	Singapore
Public Investment Fund	Saudi Arabia
Hong Kong Monetary Authority	Hong Kong
National Council for Social Security	China
Qatar Investment Authority	UAE

Quiz questions Chapter 3: Direct real estate capital markets

1 **Provide examples of where leasehold ownership is more commonplace.**
Directly on or adjacent to infrastructure facilities such as airports and ports. In these instances, the authorities seek to retain enhanced control over land and associated properties given their critical use.
2 **What is strata ownership?**
A strata ownership investor owns a specific floor or pro rata area of the reference property, with the investor(s) having obligations and rights over the common areas (e.g., lobbies), structural maintenance and other aspects such as facade signage.
3 **What are the traditional sectors of institutional real estate ownership?**
Office, retail, industrial and multifamily.
4 **Why is it important to understand estimates of the size of different real estate markets?**
These estimates help frame strategic real estate portfolio allocations around which tactical allocations and/or investor-specific considerations can be overlaid. At a geographic and/or sector level, they can also inform direct investors of market depth and liquidity profile. Finally, they also inform commercial real estate's position within broader multi-asset portfolio allocation frameworks.
5 **How is new investible stock created?**
Through development and sale and leaseback transactions.

Quiz questions Chapter 4: Analysing a direct real estate investment

1 List the key factors that drive the rental growth outlook for a commercial real estate asset.

Answer:

• Prevailing rental levels – high or low in a cyclical context
• Market vacancy
• Market supply pipeline

- Market demand, for example, financial and business services employment for offices and consumer expenditure for retail assets
- Construction cost inflation
- Contractual indexation, for example, inflation (CPI) outlooks

List the attributes that could determine either the new leasing prospects or renewal probabilities for office space.

Answer:

- Micro-location, for example, proximity to transport
- Building aesthetics
- Age profile
- Specification, for example, is it modern and does it provide a healthy work environment?
- Amenity provision
- View
- Rent level and overall affordability of occupation.
- Availability of competing space
- Lease terms
- Sustainability profile
- Macro considerations, for example, increasing prevalence of 'work-from-home'.

2 What is the difference in the calculation of gross and net initial yields?
   Answer: Gross initial yields exclude market standard acquisition costs whereas net initial yields include them in the denominator.
3 Using the illustrative direct real estate investment example cashflow in Figure 4.3, what would be the IRR dilution of a 75% increase in the budgeted capital expenditures?
   Answer: 0.4%
4 Using the illustrative direct real estate investment example cashflow in Figure 4.3, fill in the following sensitivity analysis matrix for a range of CPI growth and exit cap rate assumptions that will impact Unit 1's projected NOI.

IRR Sensitivity				CPI			
	0.0%	−2.0%	0.0%	3.0%	5.0%	7.0%
Exit Cap	6.5%					
Rate	6.0%					
	5.5%			7.6%		
	5.0%					
	4.5%					

Profit-on-Cost Sensitivity				Market Rental Growth			
	0.0%	−2.0%	0.0%	3.0%	5.0%	7.0%
Exit Cap	6.5%					
Rate	6.0%					
	5.5%			1.41x		
	5.0%					
	4.5%					

Answer:

IRR Sensitivity

7.6%	−2.0%	0.0%	CPI 3.0%	5.0%	7.0%
Exit Cap Rate 6.5%	2.4%	3.2%	4.5%	5.4%	6.4%
6.0%	3.8%	4.7%	6.0%	6.9%	7.9%
5.5%	5.5%	6.3%	7.6%	8.6%	9.6%
5.0%	7.3%	8.2%	9.5%	10.5%	11.5%
4.5%	9.4%	10.3%	11.6%	12.6%	13.7%

Profit-on-Cost Sensitivity   **Market Rental Growth**

140.9%	−2.0%	0.0%	3.0%	5.0%	7.0%
Exit Cap Rate 6.5%	1.11x	1.15x	1.22x	1.28x	1.33x
6.0%	1.19x	1.23x	1.31x	1.36x	1.43x
5.5%	1.28x	1.33x	1.41x	1.47x	1.53x
5.0%	1.39x	1.44x	1.53x	1.59x	1.67x
4.5%	1.52x	1.58x	1.68x	1.75x	1.83x

Quiz questions Chapter 5: Valuing direct real estate assets

1 What are the differences between the three market valuation approaches described in this chapter?

Answer:

• Cost approach: This simply references the build cost to either reinstate or replace an asset.
• Market comparable approach: This compares an asset's price metrics, for example, price per square metre or price per unit, to those of relevant 'peer' assets that have recently traded.
• Simple capitalisation approach: This uses relevant local market yields or cap rates to value an asset existing or potential income with reference to 'market rent' namely the level of rental income an asset would lease for at the prevailing time.

2 What factors could drive differences in risk premium and discount rates across markets?

Answer:

• Total return volatility – income and capital growth components.
• Market transparency, for example, the availability of performance series and benchmarks.
• Lease lengths and characteristics.
• Property rights/rule of law.
• Market size and depth of liquidity.
• Degree of underlying operational risk and margins.
• Risk-free rate assumptions across countries.

3 How can sustainability considerations be fed into DCF-based valuation models?

Answer: Largely taken from Sayce et al. (2022), for assets with relatively weak sustainability characteristics, the following apply:

• Reduced projected income –

  • Reduced rent and occupancy arising from falling occupier demand.
  • Increased timeframes to release space.
  • Weaker occupier profiles
  • Losses from potential changes to the permitted use of space

• Higher expected outgoings:

  • Increased operating expenses – building and utilities costs.
  • Increased capital expenditure requirements
  • Higher insurance premiums
  • Potentially higher property taxes, including 'carbon' costs.

• More elevated risk premia/discount rates:

  • Higher-income and cashflow volatility per the above
  • More intense management related to sustainability considerations.
  • Lower market liquidity from reduced investor demand
  • Less competitive debt finance terms due to lender preferences

5 Using the illustrative DCF model, what would be the change in the calculated investment worth valuation adjustment if budgeted capital expenditures had doubled?

  Answer: 25,125,332 – 24,447,278 = 678,055

6 Using the illustrative DCF model illustration example cashflow in Figure 5.3, fill in the following sensitivity analysis matrix for the investment worth calculated across a range of CPI growth and exit (year 10) cap rate assumptions.

			CPI		
	0.0%	1.0%	3.0%	5.0%	7.0%
Exit Cap Rate	6.5%				
	6.0%				
	5.5%		25.125		
	5.0%				
	4.5%				

	2287.57%	1.00%	3.00%	5.00%	7.00%
Exit Cap Rate	6.50%	22.253	22.578	22.878	23.144
	6.00%	23.407	23.746	24.056	24.329
	5.50%	24.771	25.125	25.448	25.730
	5.00%	26.408	26.781	27.119	27.411
	4.50%	28.409	28.804	29.160	29.466

Quiz questions Chapter 6: Private real estate fund structures

1 **What are the four styles of private real estate funds?**
Core, Core-Plus, Value Add, and Opportunistic
2 **What is the key difference between a Core and Core-Plus Fund?**
Leverage – A Core Fund will target leverage <40%.
3 **Identify the roles of a general partner and a limited partner.**
A general partner is the fund manager, and the limited partner is the investor/capital provider.
4 **What metrics do you require to determine an investor's net cash flow in a PREF?**

A	Gross investor cashflow	
B	Annual management pees	
C	Pre-performance fee cashflow	A - B
D	Preferred return	
E	Excess profit	C - D
F	Carried interest	E * 20%
G	Net investors' cashflow	C - F

5 **What is the typical lifespan of a closed-end fund?**
Commonly from six to seven years and up to 12 years.

Quiz questions Chapter 7: Valuing private real estate vehicles

1 **What consideration should investors factor into discount rates when valuing private real estate vehicles?**

Answer:

• Underlying real estate holdings characteristics and the risk profile that would dictate an appropriate risk premium for the asset/portfolio.
• A relevant risk-free, which could be an average of multiple countries in the case of cross-border investment holdings.
• Their effective tax rate on the investment
• Capital structure and debt financing profile.
• Liquidity profile
• Regulatory and legal stability of the structure's jurisdiction
• Investment horizon
• Any foreign currency considerations

2 **What are the calculated vehicle risk premia for the following leverage and tax assumptions assuming a market risk premium of 5%?**

a Unlevered beta = 1.1x, tax rate = 20%, LTV = 50%
b Unlevered beta = 1.25x, tax rate = 40%, LTV = 20%
c Unlevered beta = 1.5x, tax rate = 10%, LTV = 75%

Answers:
• 9.9%

- 7.2%
- 27.8%

3 **Using the illustrative DCF model in Figure 13.2, what would be the change in the calculated investment worth valuation adjustment if budgeted capital expenditures doubled?**

Answer: 12,044,399 – 11,604,301 = 440,098

4 **For a PREF with an existing portfolio that has just been expanded through a new acquisition using debt finance, calculate this PREF's NAV at the current time and in year three with and without INREV NAV adjustments using this simple balance sheet format.**

Property Assets +
Debt −
Other Net Assets/Liabilities +/−
INREV Adjustments +/−
Net Asset Value

Assumptions to use:

- Current property assets = 100 (fair market value)
- New acquisitions = 200 (fair market value)
- Capital growth = 3% p.a.
- Acquisition costs = 10%
- Debt employed = 100 (no change to this)
- Loan arrangement costs = 2%
- Other net assets/liabilities = +5 (no change to this)

Answers:

- Current: with/without adjustment 205.0 / 227.0
- Year 3: with/without adjustment 232.8 / 241.6

No INREV Adjustment	Current	Year 1	Year 2	Year 3
Property Assets	300.0	309.0	318.3	327.8
Debt	(100)	(100)	(100)	(100)
Other Net Assets/Liabilities	5	5	5	5
Net Asset Value	205.0	214.0	223.3	232.8

With INREV Adjustment	Current	Year 1	Year 2	Year 3
Property Assets	300.0	309.0	318.3	327.8
Debt	(100)	(100)	(100)	(100)

Other Net Assets/Liabilities	5	5	5	5
Capitalised Costs	22.0	17.6	13.2	8.8
Net Asset Value	227.0	231.6	236.5	241.6

Quiz questions Chapter 8: Private commercial real estate debt structures

1 **What are the key economic drivers of private commercial real estate loans?**
2 **Calculate the monthly and quarterly interest payments for the following commercial real estate loans.**

   a 50 million size, 10-year term and 5.0% interest rate
   b 10 million size, 10-year term and 6.0% interest rate
   c 50 million size, 10-year term and 5.0% interest rate
   d 50 million size, 10-year term and 4.0% interest rate

3 **For all of the above commercial real estate loans in Question 2, calculate the compound interest rates for monthly and quarterly payment periods.**
4 **Using the model for Figure 14.4 provided in the accompanying Excel file, calculate the lender's IRR or yield-to-maturity for the following two prepayment penalty scenarios:**

   a The make-whole penalty increases to 3.5 years of interest payments.
   b The lender instead is guaranteed a 1.2x profit on their capital.

Quiz questions Chapter 9: Public market structure

1 **Which market participants are behind the Chinese Wall?**
Corporate finance and compliance.
2 **List the key motivations of a sell-side analyst.**
Trading commission, research fees (post MiFID II), and (indirectly) related corporate finance fees from individual companies researched.
3 **List the key motivations of a buy-side analyst.**
Fund outperformance vs. benchmark, asset management fees (linked to size of fund and performance).
4 **Name the key variables required to assess a company's performance.**

- Asset performance
- Forecast growth rates in NAV, DPS and AFFO/EPS
- Executive remuneration
- Cost ratios (admin expenses as a % of rents)
- Gearing
- Valuation
- Share price performance.
- Shareholder structure

5 **Why do investors use a Free Float Market capitalisation weighted Index?**

Because it represents the investible amount of each company's market.

Quiz questions Chapter 10: Analysing a public real estate company

1 Why is real estate different from the other 10 equity sectors?

Real estate is asset-based

The other sectors comprise predominantly trading companies which sell products or services. Since the advent of e-commerce and globalisation, this can be 24 hours a day, 7 days a week. As a result, the key items when analysing these companies are metrics such as gross and net profit margins, operating cost ratios, stock turnover, and liquidity ratios. Real estate companies, in contrast, are either asset-based investment companies (REITs) or asset-based development companies with very lumpy sales. It should be noted that housebuilding companies are not included in the real estate sector, although certain Asian-based companies with large residential development arms are. As a result of being asset-based, standard metrics such as profit margins and liquidity ratios are not as relevant.

Income is less frequent but more predictable

Typically, a REIT receives its income four times a year (Quarter Days) from its tenant base (customers) with whom it has an ongoing contractual relationship (lease). This contrasts sharply with say, Apple, which sells a product every second of every day somewhere in the world. However, there is (normally) a greater certainty and predictability to the income of REITs due to the contractual nature of leases, compared to the sales of a trading company. To put it at its most extreme, if the management team of a REIT did not turn up to the office for a period, the tenants would still be obliged to pay their rent. In contrast, if the employees and suppliers of a supermarket, say Sainsbury's, did not turn up to work, there would be empty shelves and no sales.

There is a parallel asset pricing market

All sectors have fundamental pricing in their underlying market, which influences the share prices of companies in that sector, be it the price of oil, financial products, electricity, baked beans, or semiconductors. However, real estate is different because the investment assets a real estate company owns are valued independently of the corporate entity (the listed company). If a REIT owns a shopping centre, that centre has an investment value and can be sold separately of the REIT.

Brand value

In contrast to real estate companies, the stock that other listed companies own, be it oil, electricity, baked beans, or semiconductors, is typically held at cost and requires an operational or distribution platform to realise full value. It is the listed entity that adds value by providing the operational platform and "brand". An unbranded Apple computer or pair of Adidas/Nike trainers is worth far less than the branded version. Typically, with the rare exception of say Westfield shopping centres, real estate assets have not been branded. It is worth noting at this stage, however, that the trend is changing. Not only do the "new", "alternative" sectors such as healthcare, data centres and student accommodation have significantly higher operational requirements, but they also often have platforms which can add value, for example, Unite Students accommodation, to that of the underlying real estate.

2 **What are the 5 key accounting metrics used to value a listed real estate company?**

Net rental income
Admin expenses
Interest payable
Gross assets
Debt

3 **Why would a company choose not to use an accounting definition of earnings?**
In order to better represent the true underlying nature of their financial position. Accounting standards not designed for real estate can portray an unrealistic picture.

4 **What metrics do the 6 EPRA BRPs cover?**

Earnings
Net asset value
Net initial yield
Vacancy rate
Cost ratio
LTV

5 **What are the three key sources of data used for forecasting?**

1 **Company data**
Regulatory updates: Full-year figures, interim reports, quarterly updates,
Discretionary updates: Press releases on disposals, acquisitions, and market conditions.
Capital market days (site visits)
2 **Real estate market data**
Real estate market indices
Agents' reports.
3 **Capital market data.**
Economic data such as GDP, inflation, employment, PSBR
Changes to in-house forecasts

Quiz questions Chapter 11: Analysing public real estate debt

1 **What does the Yield-to-maturity represent and why is it used?**
Yield-to-Maturity – this reflects the total return to the investor throughout the life of the bond and is used in secondary market valuations when bond prices will differ from the issue price. This calculation allows investors to benchmark bonds against each other and an index.

2 **How does a normal bond differ from a convertible bond?**
Certain issues have an equity component in addition to the plain vanilla debt. This is known as a convertible bond, and whilst the coupon may typically be lower than for a straightforward bond, the number of investors who participate in convertibles is smaller than for straight bonds. The coupon is lower because there is theoretically some upside via the conversion rights to equity. The conversion price will be at a premium to the share price at the time of the convertible issuance.

3 **Name three ways you would analyse the size and depth of a bond market for an investor.**
Firstly, looking at the market in terms of bond duration, secondly, looking at the market in terms of credit rating and, finally, in terms of the underlying real estate sector.

4 **What are the key elements of a credit rating methodology?**

- Country analysis
- Sector analysis
- Portfolio quality
- Management Quality
- Financial Profile
- Peer Group Analysis

5 **What are the five broad categories of credit rating for bonds?**

Highest quality = AAA
Investment grade = BBB and above
Below investment grade a.k.a. Junk Bonds = BBB and below
Higher risk of default = C
In default = D

Quiz questions Chapter 12: Valuing a public real estate company

1 **What are the three-performance metrics we look at for a real estate company?**

a Cash flow generation = EPS
How much is the company expected to generate, after tax, on an annual basis?
b Cash on cash return to investors = DPS
How much can shareholders expect to receive in terms of annual cash income return?
c Value of the underlying assets = NAV per share

What is the market value of the underlying assets (less associated debt and other liabilities) regardless of share price?

2 **Define EPS**
Definition:
Earnings attributable to ordinary shareholders/weighted average number of shares during the accounting period = EPS
3 **Define NAV per share.**
Definition:
Net asset value per share (NAV) = Net assets at the end of accounting period/shares in issue at the end accounting period
4 **What valuation method is prevalent in the US?**
US – no mark-to-market of assets, so multiple of AFFO and dividend yield key metrics.
5 **What are the advantages of the dividend valuation methodology?**

Advantages
Measures true cash return to investors.

Allows direct comparison with return on underlying assets.

Can be used to compare any capital (debt or equity) market globally.

Quiz questions and answers Chapter 13: Setting a target price

1 **What is the purpose of a target price?**

A target price is set to determine the expected performance of a company's share price based on detailed forecasts, and to compare that expected performance in absolute or relative terms and produce an investment recommendation.

2 **Name two methods for determining a target price that do not involve using a weighted average.**

- This involves a long-term average discount to NAV, which is then adjusted for specific factors to arrive at a fair value discount which is applied to forecast NAVs. It is logical and simple to use and adjust.
- This methodology looks at whether a company creates positive or negative returns (i.e., above or below its weighted average cost-of-capital) and subsequently accords a premium/discount valuation to forecast NAV. It is logical but slightly more complicated to determine.

3 **What are the three methods of valuation used in a weighted average target price methodology?**

Earnings-based, dividend-based, and NAV-based.

4 **What are the four main reasons why an analyst might change a target price before the anticipated date?**

A change in global capital market assumptions.

A change in sentiment towards the sector as a whole

A change at the company level

A change at the asset level

5 **How can you determine an investment recommendation based on a target price?**

If the target price is >10% above the current share price, the recommendation is a buy. If the target price is <10% below the current share price, the recommendation is a sell.

Quiz questions and answers Chapter 14: Operational mechanics of a public real estate company

1 **What are the four main operational elements of a real estate company?**

1 Asset management

optimising the value of the real estate portfolio

2 Liability management

ensuring a prudent and optimal financing structure to ensure the real estate strategy can be implemented.

3 Corporate management

whether internally or externally managed, there is an obligation to ensure that the corporate level is administered efficiently without excessive overheads.

4 Investor relations/stock market regulations

2 **Name three types of investment analyst that would follow a listed real estate company.**

1 Financial adviser investment analyst
2 Independent investment analysts
3 Paid-for investment analysts (smaller and newer companies only)

3 **List five different investment strategies for a listed real estate company**

1 Specialist aggregator
2 Specialist major
3 Investment stylists
4 Beta plays
5 Adaptors

4 **What are the five stages of a listed real estate company's evolution?**

1 Pre-IPO
2 IPO to junior market
3 Move to main market and EPRA indices.
4 Inclusion in general equity indices
5 Leading company in sector/market

5 **What two factors would you use to determine the stage of a company's evolution?**
Shareholder type
Valuation driver

Quiz questions and answers Chapter 15: Public market applications

1 **List four ways in which the structure of a dedicated real estate securities fund might differ.**
Long-only or long-short
Actively or passively managed.
Benchmarked or non-benchmarked.
Use of relative or absolute benchmark
2 **What accounted for the growth in real estate securities funds from 2000?**
The superior performance of the underlying real estate market during that period, which, fuelled by strong occupational demand and an increase in available debt capital, saw the asset class produce a comparable return to equities with lower volatility. This superior risk-adjusted return resulted in both retail and institutional investors seeking exposure to the underlying asset class. In addition, the globalisation and securitisation that was occurring, led by the introduction of REIT legislation across the world, increased the geographic scope, absolute size, and liquidity of the investable universe. Product developers employed by asset managers were able to create collective investment vehicles that allowed this pool of capital to be harnessed in a variety of different ways.
3 **What factors determine how much a fund manager might allocate weightings away from the benchmark?**

• the level of tracking error that investors are happy to accept, as indicated in the fund prospectus.

- the size of the fund: the smaller the fund, the less likely it will be that realising individual positions will have an impact.
- whether the portfolio is constructed using the benchmark regional weightings as a starting point or whether the portfolio is assembled "bottom-up" by selecting the most attractive company valuations regardless of region
- the frequency with which the fund trades
- whether the fund is open-ended, and hence subject to inflows and redemptions from unit holders which affect portfolio weightings and activity, or close-ended.
- whether the fund has a secondary objective, such as a preferred minimum yield, liquidity, or a currency requirement.
- the number and location of staff involved in the fund. It would be unusual, for example, to take a significant underweight position in a region where most staff were located.
- the time

4 **List six other asset classes that public real estate might be combined with?**
Direct property
Internal funds
External funds
Debt
Real assets
Derivatives

5 **What are the seven ways in which you would determine whether adding public real estate to a private allocation had been successful?**
Return enhancement.
Risk-adjusted impact
Tracking error
Currency impact
Cash drag.
Risk attribution
Portfolio contribution

Index

Pages in *italics* refer to figures.

Printed in the United States
by Baker & Taylor Publisher Services